U0820679

国际住宅设计

董君 李歆怡 主编

FUTURE ARCHITECTURE/建筑未来

中国林业出版社

图书在版编目（CIP）数据

国际住宅设计——建筑未来：汉英对照 / 董君，李歆怡主编. ——北京：中国林业出版社，2011.1
ISBN 978-7-5038-6075-1

Ⅰ. ①国… Ⅱ. ①董… ②李… Ⅲ. ①住宅－建筑设计－汉、英 Ⅳ. ①TU241

中国版本图书馆CIP数据核字(2011)第008690号

主　编: 董　君　李歆怡
副主编: 刘　云　姜　野　林蜜蜜
编　写: 董　君　李歆怡　孙　亮　陈　聪　董雪妮　刘　雨
王　扬　刘　英　顾建英　丁　健　董　强　方　言
梁国文　何婷婷　黄　华　吕　强　潘　涛　任芹芹
肖　毅　张　磊　施春雪　史书明　宋　梅　谭　倩
田　宁　龙伟基　王　梅　王　斌　王立军　王　青
王震华　魏　春　徐　婷　高　震　杨　萍　江　涛
李　军　林志勇　赵晓飞　周　萍
策　划: 尚品雅图文化传播有限公司

责任编辑: 纪亮

出版: 中国林业出版社
（100009 北京西城区德内大街刘海胡同 7 号）
网址: www.cfph.com.cn
E-mail: cfphz@public.bta.net.cn
电话:（010）83223051
发行: 新华书店
印刷: 恒美印务（广州）有限公司
版次: 2011年2月第1版
印次: 2011年2月第1次
开本: 230mm×300mm
印张: 16
字数: 200千字
定价: 240.00元

序

2010年上海世博会的成功举办向全世界展现了一个崭新的中国、强盛的中国、有着悠久历史与灿烂文化的中国，同时，世界其他国家也通过建筑、展览、文化交流等形式向中国传递着它们的生活方式和建筑理念。经济的高速发展，信息一体化的加速进程，成就了世界各地建筑行业的空前繁荣。在过去的一年中，遍布各地的无数新的建筑产生，无数优秀的住宅设计精品涌现。在期待中，《国际住宅设计》一书即将付梓面世了，为本书出版作序，我深表荣幸。

《国际住宅设计》精心编辑了30套精美的住宅建筑及室内设计的最新案例。案例包含了亚洲、美洲、欧洲等诸多地区。极具异域风情和多元文化气息的空间形式，相信会带给读者截然不同的感受和思考。全部案例均为世界顶尖建筑公司及一线设计师近期的优秀作品，该书以图文并茂的形式，通过美轮美奂的图片将每处建筑、室内、周边景观以及局部设计风格处理都做出细致、系统的展示，是一本使用价值颇高的参考书。每套案例都配有详细的设计说明和设计亮点陈述。 其所引入并展现的许多优秀的让人耳目一新的住宅理念，无论其杰出的创造性以及一流的功能品质都令人信服。

建筑行业是一个大家庭。行业的成长，水平的提高仰赖于行业方方面面的共同努力和真诚合作。相信《国际住宅设计》对中国优秀设计企业，设计精英的服务和支持，将为广大同行提供一个良好的专业交流平台。这些具有较高设计水准的国际案例，在一定程度上可以成为业内住宅建筑设计潮流的风向标。整套书注重品质与实用的统一，专业说明和高精度图片并举，是一套集欣赏性、可读性和实用性于一身的优秀力作。

本书不仅是建筑爱好者和私宅业主的良师益友，更是耕耘于住宅设计领域的建筑师和设计师一泓清新的、重要的灵感源泉，甚至对于城市建设公司、房地产开发商、城市建设的决策者，以及我国各地目前正蓬勃发展的城市住宅建设有着直接和重要的借鉴作用。

2011年2月

前言

本书荟萃了30套最新完成的、遍布世界各地的优秀住宅设计实例，它们在建筑式样和功能设计上都堪称精品。这些项目由全世界一批最引人注目的、最有创造力的建筑公司和建筑师团队设计制作而成，其中包括查尔斯·罗斯建筑公司、DCPP建筑设计公司、EASTERN设计组、HOJO建筑结构研究院、XTEN设计团队以及奥比·G·鲍曼以及约瑟夫·N·比昂多建筑公司。本书收集的案例无论是在设计上还是在用材和建造方法上都是别出心裁的，每一所住宅均附有简明的案例分析，着重叙述了住宅的设计、技术处理以及材料、承造者的情况和居住环境等，并配有当今最棒的建筑摄影师所拍摄的精美的照片。这些照片对空间整体布局和细节作了全方位的展示，力求使读者更透彻地了解设计理念，从而获得直观的参考价值。每个案例均附有各层平面图、剖面图及细部节点图并陈述了设计的最大特点。

结构数据与经济指标在便于方案比较的同时也为设计者提供了有价值的信息。笔者相信通过本书的阅读可以帮助读者了解世界当下流行的先进的设计理念，感受未来设计发展的脉络。富有创造精神的大胆设计和一丝不苟的实践使得这些带有浓郁“实验”色彩的“前卫”作品被实现在了世界各地形形色色的住宅建筑类型上，从都市寓所到郊区住宅、从滨水居所到乡间家园、从常年居住的住宅到季节性或周末度假的小屋，各种规模、各种品质皆各具特色。

精心构思的设计方案，先进的设计理念，精致的施工工艺和最终呈现的功能与品质，使得本书适应人群广，参考价值高。不仅是为建筑、园林和规划专业的学生、设计师编写的，对于从事城市园林和规划的管理者、建筑及室内装修行业的专业从业者，甚至准备居室装修的大众也能从中寻找装修灵感，开阔思路。笔者强烈推荐此书给建筑及相关专业的同行们，以期提升其设计水平，更深入地了解与世界顶尖设计的差距，更深入地研究住宅建筑类型与生活方式之间的密切关系，延续我国历史和建筑文脉，在中国地理文化基础上大胆创造，共同努力提升中国未来住宅建筑的品质。

董君

2010年12月

Project: A House Made of Two
Architects: Akio Nakasa
Location: Kanagawa, Japan
Building area: 105.88 sqm
Photographer: Toshiyuki Yano / Nacasa & Partners

This is a project of one family living in two houses built slightly apart from one another. One great volume of a house which covers the entire plot was supposed first. Then this volume was carved in a curve in three segments and two volumes at the both ends were built as two houses on the site. The wood structures and finishing materials of two houses are standardized to emphasis the relations of two volumes being originally from "one great volume" and that they are one though apart. Carved volume in the middle became a courtyard leading to the approach to two houses. Having another house for movies at night, overnight guests or work at home allows the inhabitants the liberty to do what they please at a time they want. This distance between two houses may be just right for inviting grandparents to move in someday or providing privacy to children at puberty. The purpose of building two houses slightly apart is not the same as that of having a villa at a resort or a farm household having a main building and a barn. This project is an experiment for allowing more life style options in the life based in the urban area.

A House Ma

E OF TWO

004 项目名称: **A House Made of Two**
建筑设计: **Akio Nakasa**
项目地址: 日本神奈川县
建筑面积: **105.88** 平方米
图片: **Toshiyuki Yano/ Nacasa & Partners**

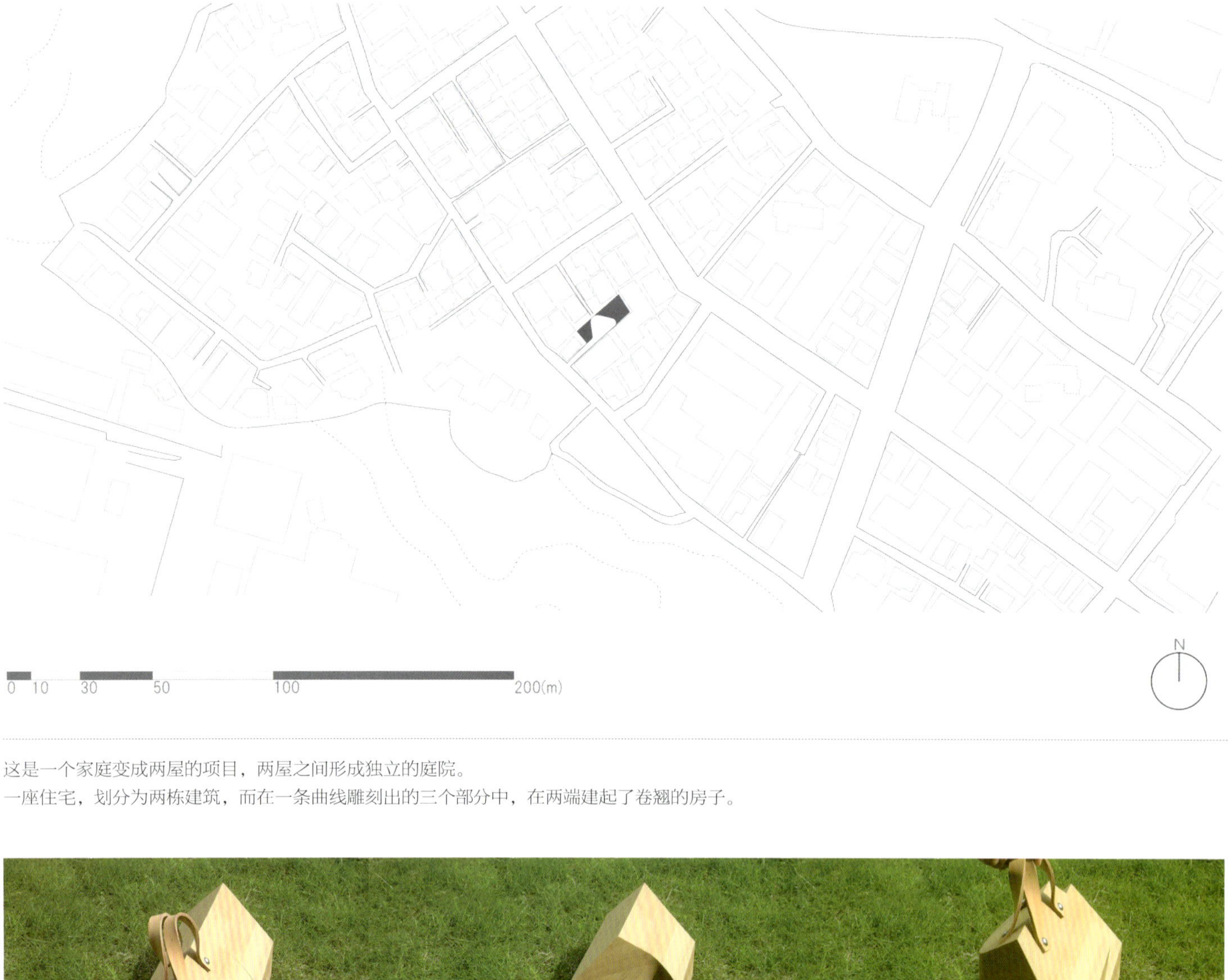

这是一个家庭变成两屋的项目，两屋之间形成独立的庭院。
一座住宅，划分为两栋建筑，而在一条曲线雕刻出的三个部分中，在两端建起了卷翘的房子。

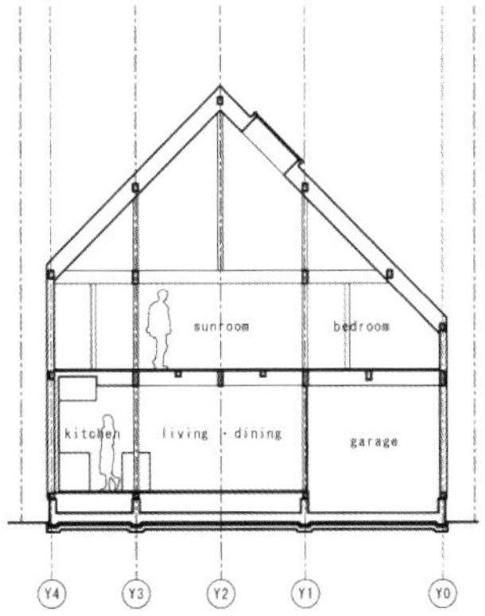

section 1

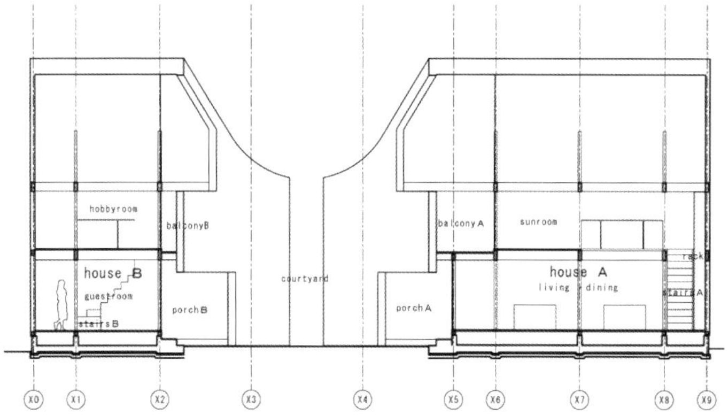

section 2

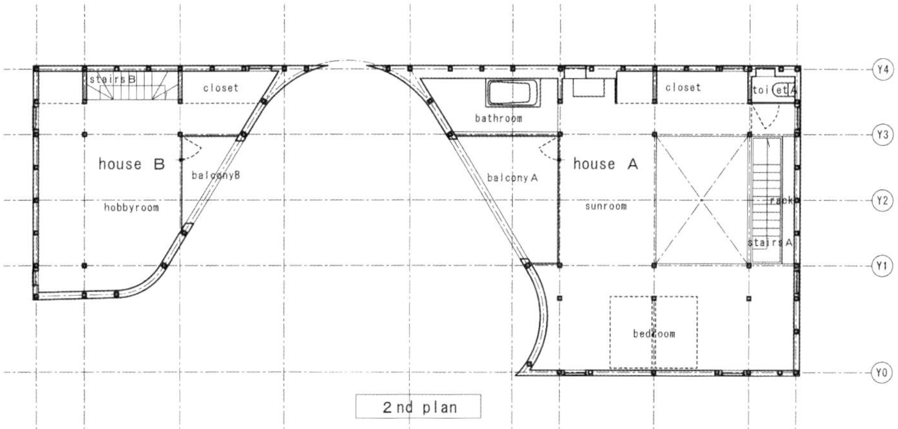

2nd plan

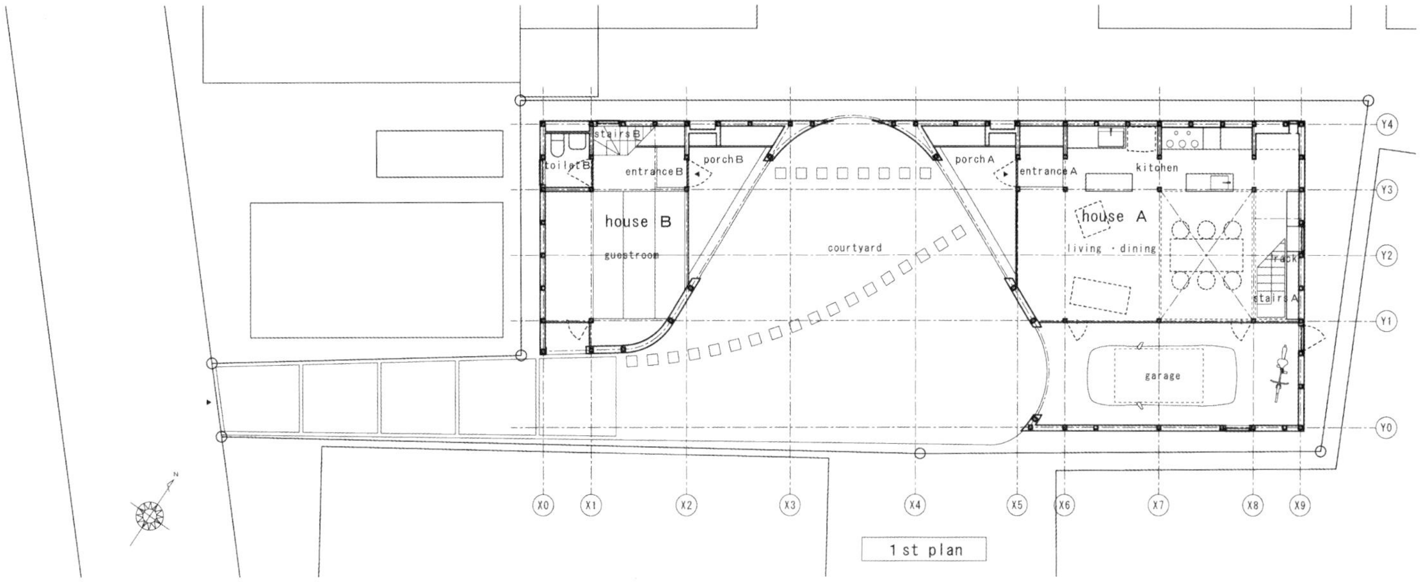

1st plan

在这个结构中，重点的地方有两个，即：“木材结构”和“两个宅院材料的统一”。从项目整体的关系来看，他们是分开的。中间划分出的庭院，是连接的方式。可以在晚上放一场电影、通宵待客或在家工作，这个房子允许我们自由地做我们想干的事情。这两栋房子之间的距离可以让房子的主人与住在一起的父母或是子女保持隐私，拥有独立的空间。

建设这套住宅的初衷稍微有些与众不同，因为它还要兼顾度假的需要。通过这个项目，让生活在市区的人们在生活方式上有了更好的选择。V

Project: Alligator
Architects: Buildingstudio
Location: New Orleans, LA / USA
Exterior images: Will Crocker, New Orleans, LA
Interior images: Undine Prohl, Santa Monica, LA

This affordable home arose out of the post-Katrina re-housing effort in New Orleans for an inner city neighborhood in dour need of new places to live. buildingstudio was working with an affluent client in Boulder, CO who voiced great concern for the lack of effort being made in New Orleans after the storm. As result we asked if they'd be willing to contribute toward an affordable home for a Katrina refugee. Not only did the clients generously give their own money, they invited their friends and colleagues to participate towards the cost of constructing an affordable home. The total sum contributed was $50,000.00. This generous contribution allows Neighborhood Housing Services, who promotes and markets low-cost properties in economically-strapped neighborhoods of New Orleans, to offer the house at a vastly reduced rate. buildingstudio, contributed its full design and coordination services as well.

The project is located in Central City, an impoverished inner-city neighborhood situated between the CBD and the prosperous Uptown area. Though Central City's infrastructure was severely deteriorating before the storm and it has one of the highest per capita of drug-related murder rates in the country, a resilient community fortitude still abounds. Infill housing is actively taking place on several fronts. Non-profits and faith-based groups have built several dozen homes for new residents hoping to move back. Most structures are either panelized or modular with little concern for design.
With an unusually narrow lot width of 19-ft, a maximum 13-ft wide scheme was all that was allowed. This resulted in an 9600 SF two-bedroom, bath-and-a-half scheme. Nicknamed "Alligator", because of its "open-mouth" profile, its based on a very common traditional typology in the New Orleans area— the shotgun. This layout is where all rooms align in a single row, front to back, in order to fit the long narrow lots typical of the City. In this design though, to overcome the lack of privacy for the middle bedroom that a shotgun typically demands, a set of rolling doors permits privacy and allows passage to the rear bedroom and bath. In older neighborhoods like Central City, the front stoop is key since so much community involvement takes place on the street. It's the common gathering place where neighbors meet and entertain each other. For that purpose, metal grate steps stretch across the front of the house for casual sitting while visiting with neighbors. The street facade is clad in translucent white insulated plastic panels on each side of the stud structure. Between these panels, fluorescent fixtures provide indirect lighting both for the exterior and interior living room. Off the rear main bedroom there is a narrow rear porch the width of the home. While considered low-cost, the budget still allowed for engineered hardwood flooring and ceramic tile in the bathrooms. The home, meeting hurricane resistance requirements, has insulated windows with an efficient thermal envelope to lower utility bills. Exterior walls and roof are clad in factory-painted, preformed metal siding for ease of maintenance and long-term durability.

Alligator

项目名称：鳄鱼
建筑设计：**Buildingstudio**
项目地址：美国新奥尔良
外部照片：**Will Crocker,** 新奥尔良，洛杉矶
室内照片：**Undine Prohl,** 圣莫尼卡，洛杉矶

north elevation

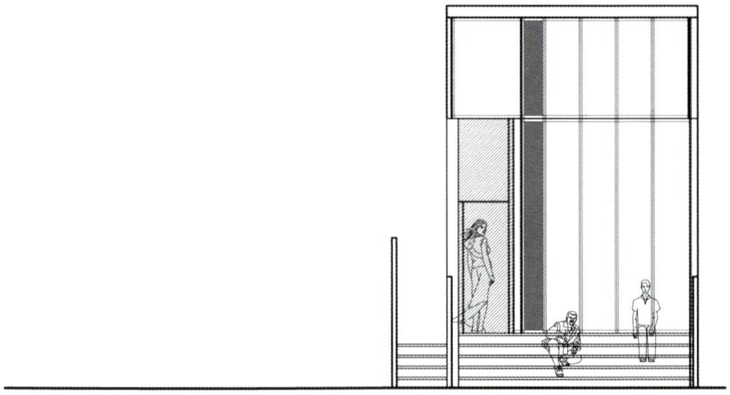

east elevation

该项目位于环市的一个贫穷内陆城市之中，毗邻中央商务区和繁华小区。虽然市中心的房屋受到暴风雨的侵袭和该地区高犯罪率的威胁。但总的来讲，多数都在很多方面采取了措施，数十家非营利机构和宗教团体正在为居民的回迁工作而努力，让他们建立新的希望。

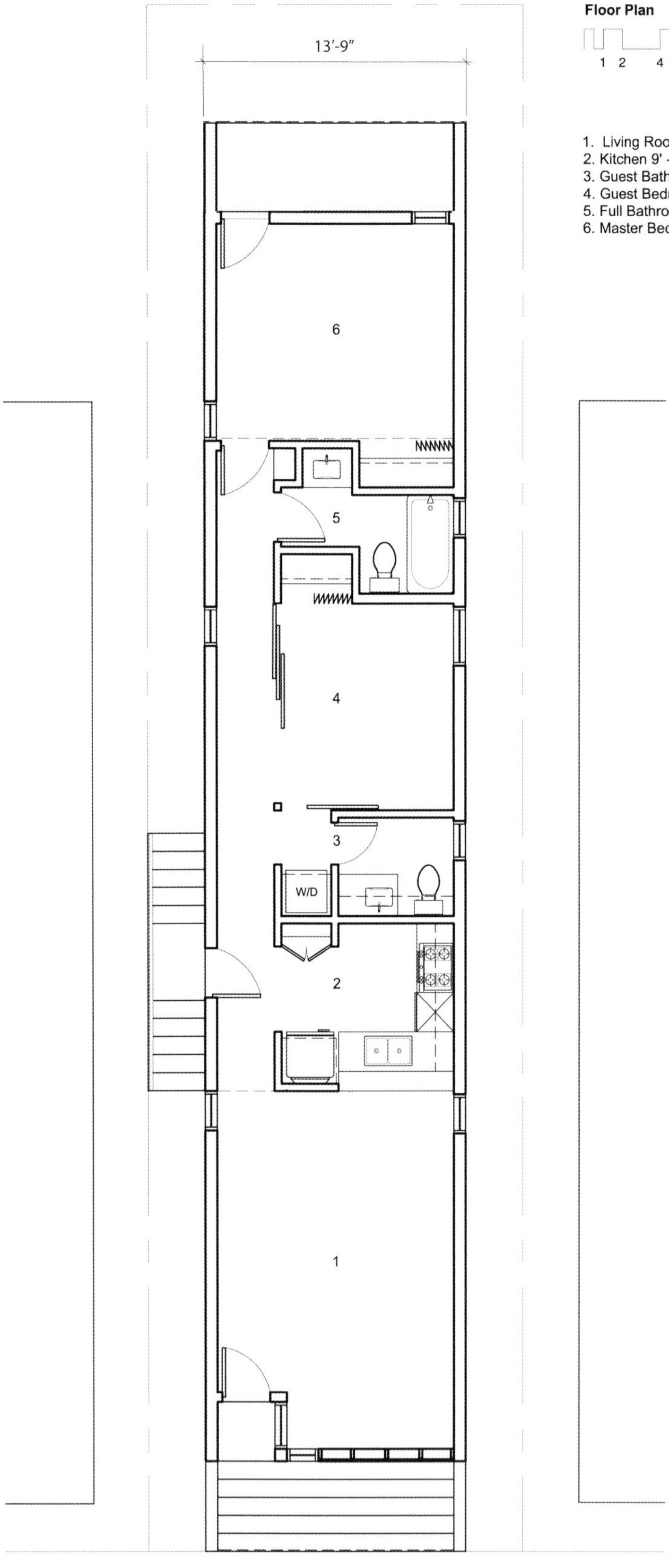

Floor Plan

1. Living Room 18' - 2" X 12' - 6"
2. Kitchen 9' - 6" X 8' - 6"
3. Guest Bathroom 6' - 2" X 5' - 0"
4. Guest Bedroom 11' - 0" X 9' - 2"
5. Full Bathroom 9' - 2" X 5' - 0"
6. Master Bedroom 11' - 0" X 12' - 6"

项目坐落在一个19英尺长的狭窄地段，只允许容纳13英尺的宽度。这个空间被规划为两间卧室、浴室和半规划区。绰号“鳄鱼”，因为其“敞开的大嘴”的独创设计。它的理念来源于新奥尔良一种常见传统类型的猎枪，这种房屋布局特点是单行排列，一直从前到后。为适应狭长的格局类型，虽然在此设计需要克服卧室中间缺乏隐私的弊端，允许“动态隐私”，也就是当有人通向后面的卧室和浴室时依然能保持私密性。例如市区的居民点，关键在于门廊，临街的社区通过前门廊，这块场地为居民提供聚会娱乐之需。

在这样一栋房屋前面坐着，身心得到舒展，还可以用来休闲、与邻居聊天。建筑临街的一面，是白色、透光绝缘的塑料板材，上面的每一个螺栓，组成各个小组件，其间有荧光灯，提供了外部和内部客厅的间接照明，虽然有降低成本的考虑，但仍然在浴室中选用硬木地板和瓷砖。这是为考虑到房屋遭遇飓风的情况，采用高效保温绝缘密封窗口，以节省能耗。外墙屋顶都进行了烤漆处理，通过预先维修和长期耐久材料来缓解金属壁板的承重压力。V

Project: Edge House
Architect:Jarmund / Vigsn's AS Architects MNAL ect.
Location: Kolbotn, Norway
Building area: 160 sqm
Photographer:Ivan Brodey, Nils Petter Dale

The Edge House is located at Kolbotn, a suburb south of Oslo.

The client, a young couple, asked for a spectacular house on a limited budget. They had purchased a challenging site with 8 meters height difference from the access road to a plateau, and they wanted a house that "looked like you could shoot a James Bond movie in it".

To save the plateau, the building was pushed towards the eastern perimeter of the site, suspended above the slope on slender steel columns. The entrance stair rises along the slope through the house up to the plateau. This strategy avoids costly blasting, hiding the technical connections in the stair. It saves at the same time the existing characteristics of the site, creating a dramatic interplay between volume and site. The entrance condition and the experience from the inside attempts to underline this interplay.

The compact interior is horizontally organized around the cut for the entrance stair. Bedrooms and bathrooms are effectively organized along a corridor.

The main structure is steel, with a polished concrete floor slab. The interior is clad in Birch plywood, the exterior in naturally colored fiber cement boards.

Edge House

项目名称: **Edge House**
建筑设计: **Jarmund / Vigsn's AS Architects MNAL ect.**
项目地址: **Kolbotn**/挪威
建筑面积: **160** 平方米
图片: **Ivan Brodey, Nils Petter Dale**

这座别墅位于挪威首都奥斯陆南部郊区。业主是一对年轻夫妇，他们对设计师提出了一个有意思的要求：要这座房子“看上去能在里面拍詹姆士邦德的电影”。这样有趣的要求似乎更能唤起设计师的想像力和创作激情。但从场地基底本身来看，便已经很具有挑战性：基底面积狭小，从入口到山体平台处有8m的高差。要在这样的环境下建造住宅已十分不易，而且又要在有限的预算下造出一座符合业主要求，或者说是“奇异”的建筑，其困难可想而知。

别墅的卧室朝向山体平台空旷处，由周围山石树木围合起来，形成一处私密领地，卧室通透的落地玻璃窗，让山色尽收眼底。起居空间面向小镇街道布置，室内公共活动与窗外市井气息贯通于此，而巨大的高差使建筑与市镇保留了足够的距离感。辅助空间位于街道一侧，狭窄开窗使建筑对外展现了较为内向的特征。建筑“面山”与“面市”的两种表情表达出了对于静谧与喧嚣的不同态度。入口设置同时保留了场地的特征，激发出了体量和场地之间神奇的相互作用，“向上”和“穿越”两种感受同时出现，即“爬山”与“进屋”两个动作一同实现。由地面入口处通过楼梯到达平台，再曲折进入建筑流线的设计，也使空间层次更为丰富。两建筑的平面，由平行于街道的走廊贯通，卧室和浴室沿走廊两侧组织，起居室位于一端，两部分在建筑入口处开始分隔。V

Project: A Vision of Temporary Living
Design: Werner aisslinger\tina bunyaprasit
Wood structure: Markus lobis

Sustainable nomadic house:
made entirely of local wood, the building provides 47 m^2 of living space with a minimal CO_2 footprint: local suppliers and local crafts using local long-lasting and recyclable materials manufactured with the precision and care of tyrolese handwork. The FINCUBE is a materialized vision of a small housing unit with a long lifecycle. It can easily be dismantled and rebuilt on a new site, and even more important for nature hideaways: it requires minimum soil sealing - just 2 m^2 hat are easily renatured after the FINCUBE is moved to another location.

Long-lasting design:
The design is minimal, material-orientated, and in close touch with nature - the wooden space with a 360-degree triple glazing is furnished with a second facade layer, producing shade and giving the building a unique overall mushroom-like monoshape. The horizontal ledges give privacy to the FINCUBE and embed the building into forests, meadows, mountain sides or any nature resorts. The combination of long-lasting design and the option of changing its location after a while make the FINCUBE a flexible home or hideaway and a lifetime companion.

Technology & space:
Technology wise the FINCUBE is a smart house - all vital house-functions are controlled by a central touchpanel.
The supporting structure is made of local larch and the interior is a combination of larch & stone-pine. The 3m-high space is organized in a helical structure: the entrance area blends into a generous open kitchen with an adjacent sofa living space, around the corner one enters the bedroom and further down is the spacious bathroom.

TE

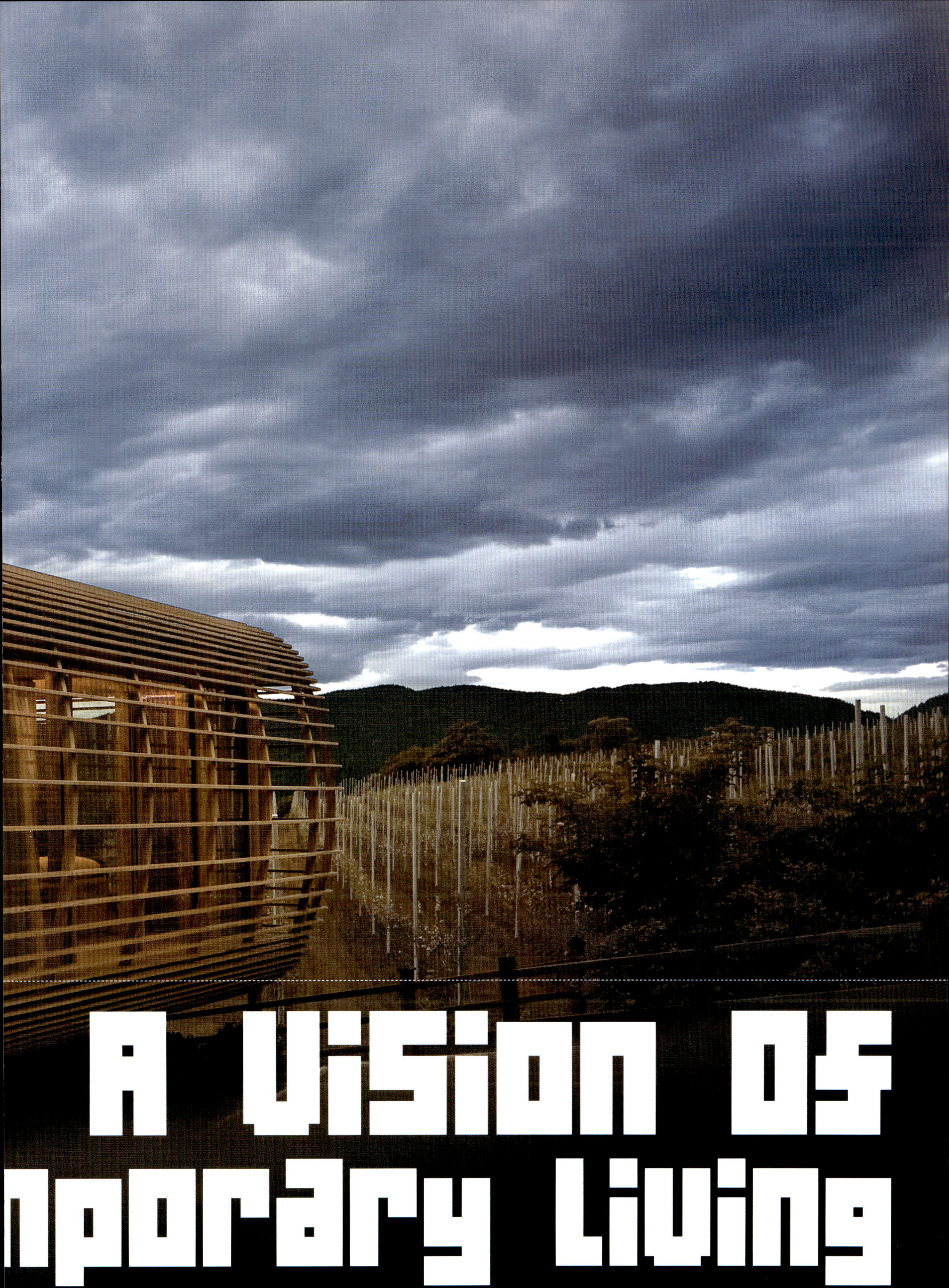

A Vision of
nporary Living

项目名称：临时生活住宅
建筑设计：**Werner aisslinger、 Tina bunyaprasit**
木质结构：**Markus lobis**

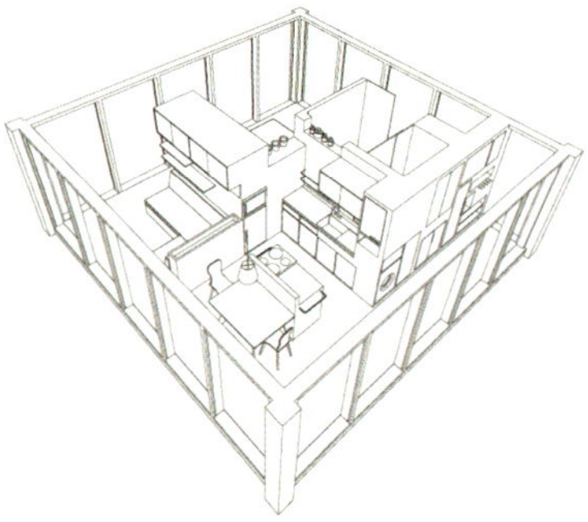

可游牧的房子：建筑材料完全用当地木材，建筑面积为$47m^2$的空间，却有着最小的二氧化碳排放。房子由本地供应商和手工艺人利用当地长期持久和可循环再造勺材料制造而成。FINCUBE这个项目力求以一个小的住宅来实现长使用周期的目标。它可以很容易拆除并重建，更重要的是和自然融为一体：项目与少量混凝土攵结合很容易被移动到另一个位置。

长期持久的设计：最小的设计，以材料为导向，且亲近自然——以360度三层玻璃木的空间装饰层与第二面层连接，进行遮阳，并建设成一个独特的整体蘑菇状。水平边界更好的让项目嵌入森林、草原、山地或任何性质的度假胜地。经过长期的设计和改变其位置，使FINCUBE融入家庭或灵活地组合变为终生的伴侣。

技术与空间：FINCUBE住宅是一个智能的房子——房子的智能装置均能由中央触摸屏控制。建筑的主要支撑结构是由当地落叶松和石松构成。建筑3m高的空间在一个有组织的螺旋结构中展开：开放式的厨房成为一个开阔的入口区，这里融合了生活空间，在沙发的拐角处进入卧室，然后则是宽敞的浴室。

PIPER
Der aufrechte Gang

Project: A Villa in Melbourne
Architects: McBride Charles - Ryan
Location: Mornington Peninsula, Australia
Building area: 258 sqm
Photographer: Courtesy of MCR

The villa area of 258 m^2, was a multi-angle of the broken line, the main cast with concrete made of black metal roof extends from the building at the top of a wall part of the building standing on this beach in Melbourne.

With the precise calculation mathematician, architect obsessed with building the relationship between structure and materials and matching to create incredible images to the building. In the design process, CAD CAD played a very important role, it will construct some very complex spatial and spatial links between analyze the issues and the establishment of a visual architectural model. The most common architectural form is used mostly as a right angle turn, the formation of most of space is a cube, so the connection between space and is relatively simple. In contrast, our "Klein bottle" really is a work beyond the age.

Villa owner is a young Australian couple, and villas in the specific address is only 1.5 hours drive from Melbourne's Mornington Peninsula. Tea in a beautiful forest which, building straight-line distance from the nearest beach is only a few meters. The start, MCR just want this design is defined as a pleasant vacation trip of tea trees. They chose "Klein bottle" a highly visual impact such a mathematical model to Sheji building exterior, while inside the building can not ignore the functional requirements, after all, this is a real use to residential. Based on this consideration, in their design, construction such as origami folding generally bent to form a special appearance.

When Klein bottle surface into the internal space of time, designers have abandoned the heavy black, instead of some bright color, such as in a lot of space is a large bright red. Interior space due to the impact of architectural form, divided into several functional areas high and low. Houses around a central courtyard, by a large step to connect all the luxury level, creating a near or far distance. Interior area of a large living and dining is part of all the members here for rest and meals. Other features include space for bedroom, laundry room, recreation room and bathroom.

If the MCR has designed the building beyond the outward features of the times, while the internal design is still using the traditional approach to treatment. Villa's internal review is still able to see many significant historic features of the interior, some furniture and decoration style and even a little bit of Melbourne with 50 years of last century building feel.

When we refer to when the building plan, it is not difficult to find in this form of design of buildings and surrounding environment is so consistent. Level surface and turn ups and downs, like patchwork tea trees. The appearance of complex rotation, brightest materials and traditional cement sheet beach house a great difference. And building a strong geometric shape different from Villa East, North have an open deck. Preservative wood deck and metal facade texture is entirely different. But in and around the plant but has a good echo. Looming on the beach of black stones, like the meteorite as full of mystery, but Villa are a huge black metal plate under the deck against the background distribution of the light.

A Villa

melbourne

项目名称: **A Villa in Melbourne**
建筑设计: **McBride Charles - Ryan**
项目地址: 澳大利亚莫宁顿半岛
建筑面积: **258** 平方米
图片: **Courtesy of MCR**

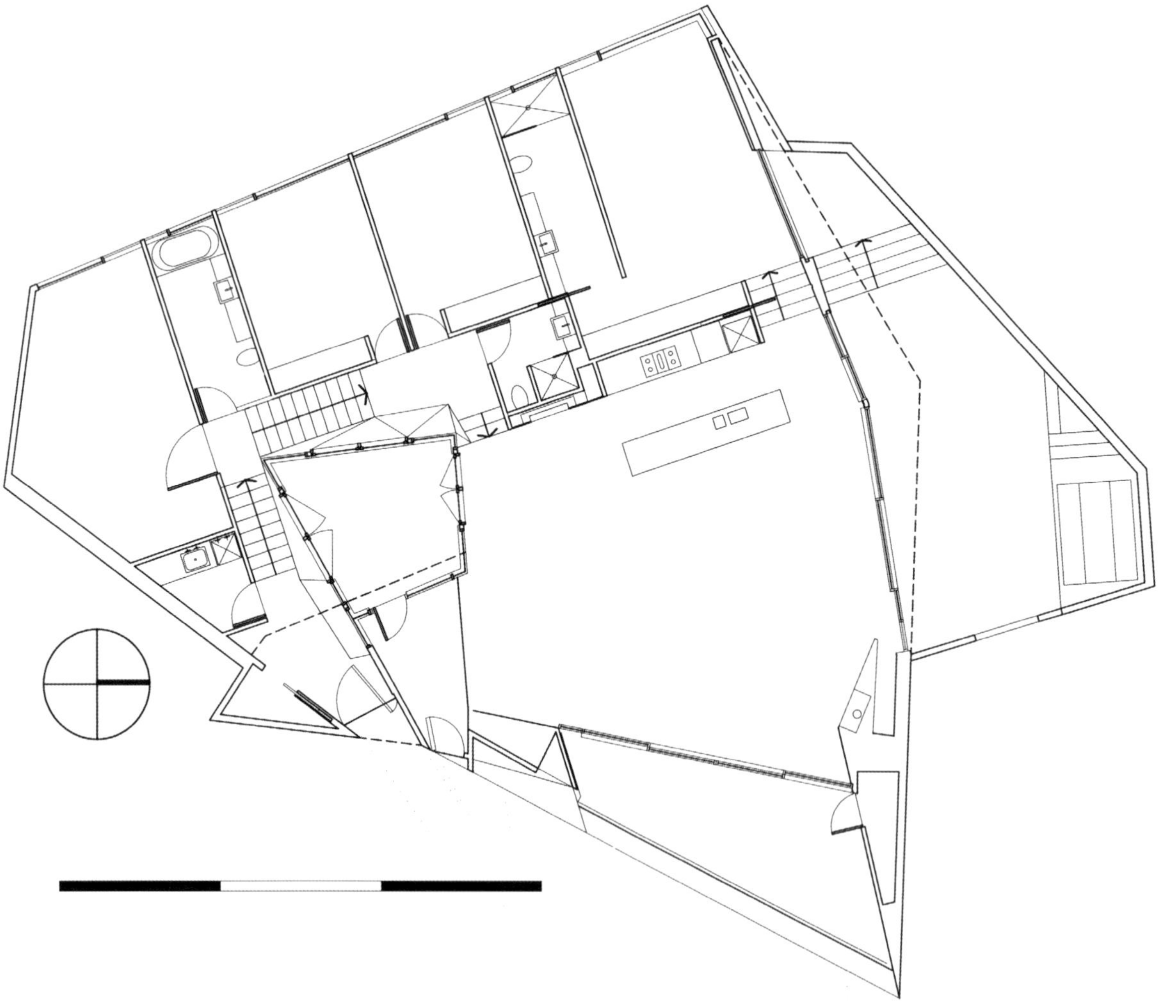

整个别墅面积为258m^2，呈一个多角的折线形，主要用混凝土浇注而成，黑色的金属屋顶从建筑顶部延伸成为外墙的一部分，建筑就这样伫立在墨尔本的海滩上。借助数学家的精密计算，建筑师醉心于建筑结构和材料之间的关系和匹配，创造出令人难以置信的建筑形象来。在整个设计过程中，CAD辅助设计担当了非常重要的角色，它将一些非常复杂的空间建构和空间之间的联系分析得清清楚楚，并建立了直观的建筑模型。

我们常见的建筑形态大多是用直角作为转折，形成的空间也大多是一个立方体，这样的空间和之间的连接是比较简单的。相比之下，我们的“克莱因瓶”真的是一件超越时代的作品。别墅的主人是一对年轻的澳大利亚夫妇，而别墅的具体地址是离墨尔本只有1.5小时车程的莫宁顿半岛。在一片美丽的茶树林里面，建筑离最近的海滩直线距离只有几米。一开始，MCR就想把这个设计定义为一次愉快的茶树林度假之旅。他们选择了“克莱因瓶”这样一个极具视觉冲击力的数学模型，来设计建筑的外观，同时又不能忽略建筑内部的功能要求，毕竟这是一个要真正使用的住宅项目。基于这样的考虑，在他们的设计中，建筑像折纸一般被折叠弯曲，形成一个特别的外观。

当曲面转入内部空间的时候，设计师便抛弃了沉重的黑色，取而代之的是一些明亮的颜色，比如在很多空间里是大面积鲜艳的红色。室内空间由于受到建筑造型的影响，分成了高低不同的几个功能区域。房子围绕的中央庭院，由一个大型的豪华台阶连接所有的水平面，创造一种忽近忽远的距离感。内部空间占地面积较大的是起居和餐厅部分，这里可供所有成员休息和用餐。其他的功能空间还包括卧室、洗衣房、娱乐室和卫生间等。

如果说MCR设计的这个建筑有着超越时代的外形特征，而内部的设计却依然是用传统的手法去处理的。回顾别墅的内部，依然能看到许多有明显历史特征的室内设计，一些家具和装饰的风格甚至带有一点点墨尔本上个世纪50年代建筑的感觉。

当我们参照这个建筑的平面图时，就不难发现用这种形式设计的建筑和周围的环境是那么的一致。高低转折的曲面和起伏的茶树林一样错落有致。复杂的旋转外观、亮眼材质与传统水泥板料的海滩小屋大异其趣。和建筑强烈的几何造型不同，别墅东、北各有一个开阔的甲板。防腐木的甲板和建筑立面的金属板质感完全不同，但是和周围的植物却有着很好的呼应。海滩上若隐若现的黑色石头，像陨石一样充满神秘感，而别墅的黑色金属板也在巨大甲板的映衬下散发着光芒。

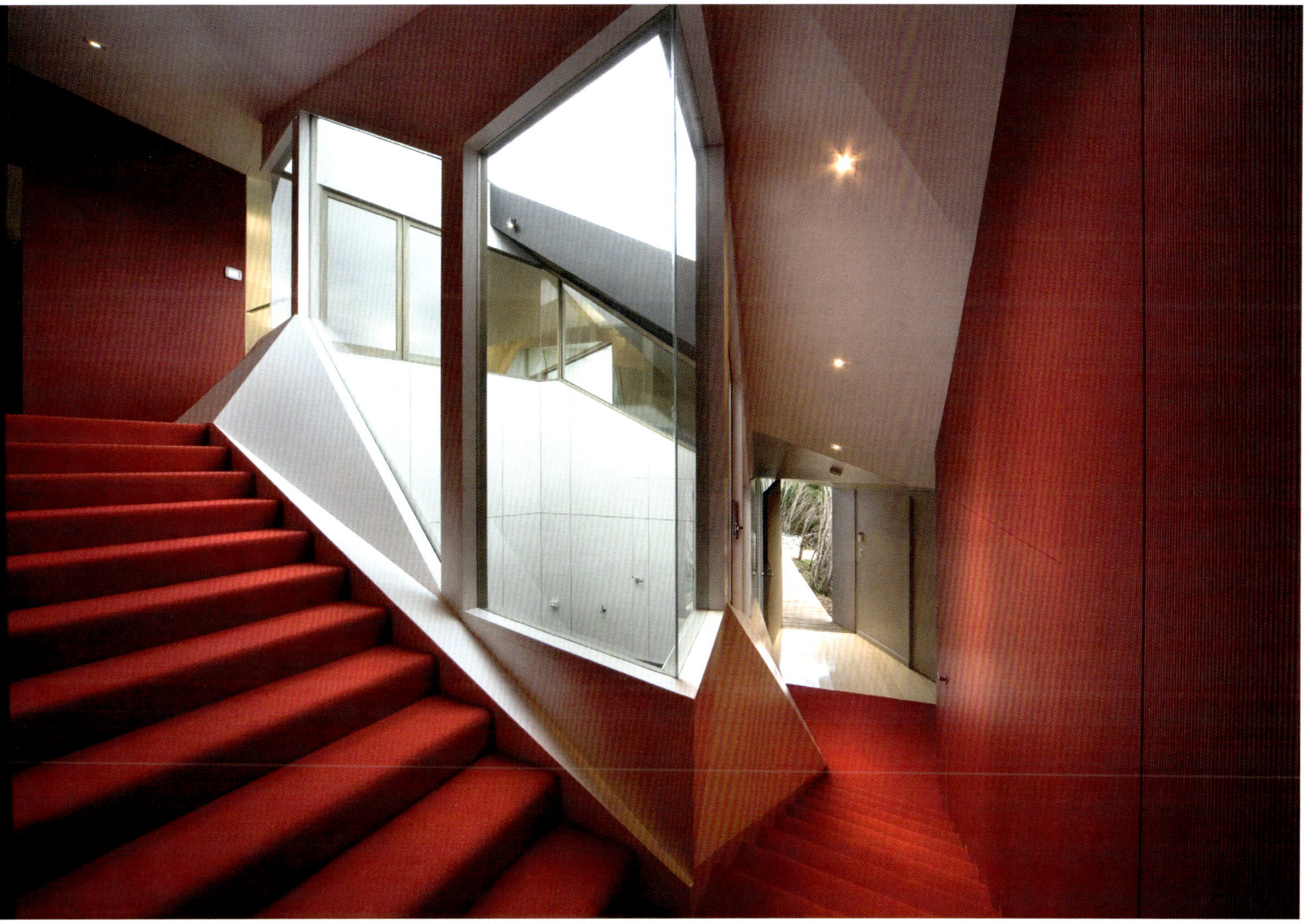

Project: Forest Town Long House
Architect: Pierre Swanepoel
Location: Johannesburg, South Africa
Photographer: Tristan McLaren & studioMAS

The Forest Town Long House is one of a kind. It responds to the many layers that make up South African living, such as safety, comfort and privacy to create a contemporary urban home. Unlike most current architecture in Johannesburg, the building itself functions as the boundary wall, containing three levels behind its front elevation . The basement level is flush with street and contains a cellar, workshop and an office with a private patio on either side of a garage. The next level contains the main living and dining areas as well as a movie theatre and family room. To one side, the open plan kitchen can be found that has direct access to an extensive 5m wide by 30m long vegetable garden. From this level, one can either access the study on basement level or, once the 12m long glass sliding door is open, the densely grown garden and swimming pool beyond. Two mezzanines overlook the double volume entertainment area and living room; a reading room and a study. From the study landing, two bedroom wings stretch out. The main bedroom with en suite bathroom and dressing room can be found to the west, where the rest of the bedrooms are found to the east. While the northern front elevation consists of smaller windows facing the street, the southern front elevation is almost completely glazed, framing the natural canvas of the magnificent garden. Completely hidden from view the back of the house opens onto a forest of massive trees and dense ground cover.

Forest T

un Long House

项目名称: **Forest Town Long House**
建筑设计: **Pierre Swanepoel**
项目地址: 南非约翰内斯堡
图片: **Tristan McLaren & studioMAS**

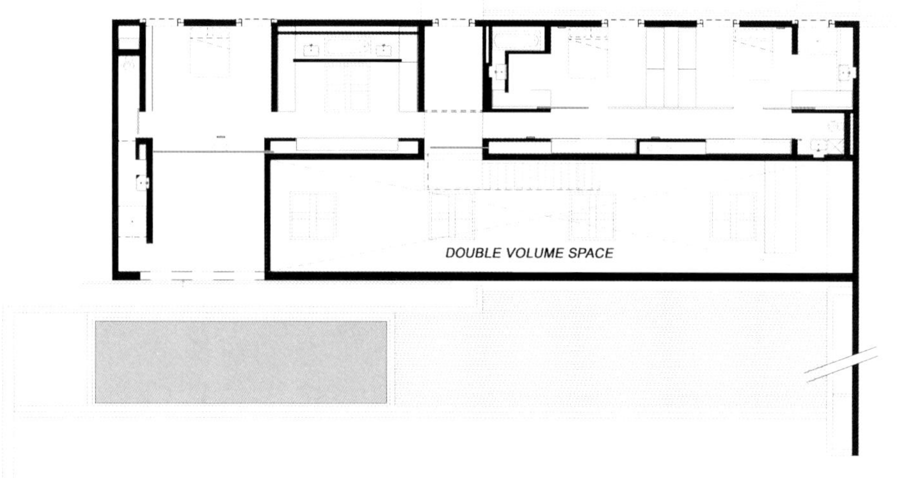

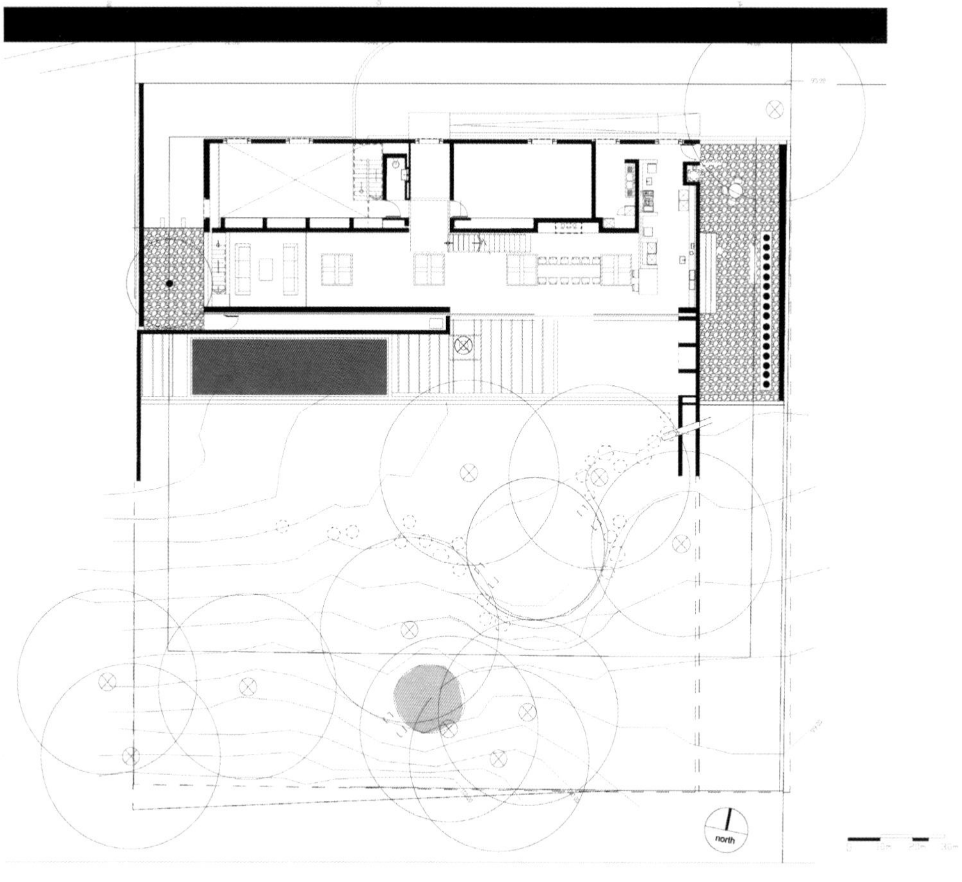

森林城条形建筑是一种形态。它的多层建筑响应了在南非生活所需要的安全、舒适和私密性，构建了一个现代的都市建筑。与大多数在约翰内斯堡当前建筑体系结构不同，建筑自身的功能为界墙，其背后包含三层。地下一层与街道相互连接，并包含一个地窖、车房和一个带有车库的单侧面私人庭院办公室。另一层包含主要的生活区和用餐区以及电影放映和家庭公共区。另外一方面，开放式厨房可以俯览到30m长，5m宽的宽阔的菜园。从这层，人们可以去往地下一层，12m长的玻璃推拉门打开时，生长茂密的花园超越游泳池的体量。两个双重娱乐区和客厅，一个阅读室用来学习。

从研究的目的来看，两间卧室向两侧延伸。带浴室和更衣室的主卧室可以看到西面的风光，其余的卧室能看到东面风景。而北立面组成的小窗户临街，南立面几乎完全透明，宏伟的花园像块天然画布，房子后面的大规模景色、树木和茂密的森林尽收眼底。V

Project: Jungfrugatan Apartment
Location: Hagagatan
Building area: 55.88 sqm

Great contrast can often bring unimaginable harmony, black and white classic brewing numerous, price of the house to square once again pushed the ultimate classic.
Metal Ware's succinct style, fine polishing, eliminating the complex concoction of cold over the Gaza Strip's only pure with modern feeling. Black out the whole cabinet from time to time in the hinge to expose the white lines, highly decorative sense, at the same time without losing the overall stringent.
Light of reasonable intervention for black and white interpretation of the symphony and thus to reconcile with gray, dim white light hang around during the Pan naturally lead people into the dream of the future state.
Natural soft tones of wood flooring from a psychological dependence gives a soft, medium, and the intense contrast of black and white, double the comfort.

The And

ltimate Black
White Beauty

项目名称: **Jungfrugatan Apartment**
项目地址: **Hagagatan**
建筑面积: **55.88** 平方米

灯光的合理介入，让黑与白演绎的交响曲因而有了装饰感，同时又不失整体的严谨。灯光的合理介入，让黑与白演绎的交响曲因而有了灰色的调和，朦胧的白色泛光流连其间，自然而然地带领人们进入了梦幻的未来境界。自然柔和色调的实木地板从心理上给人以柔软的依靠，中和了黑与白的激烈对比，舒适感倍增。

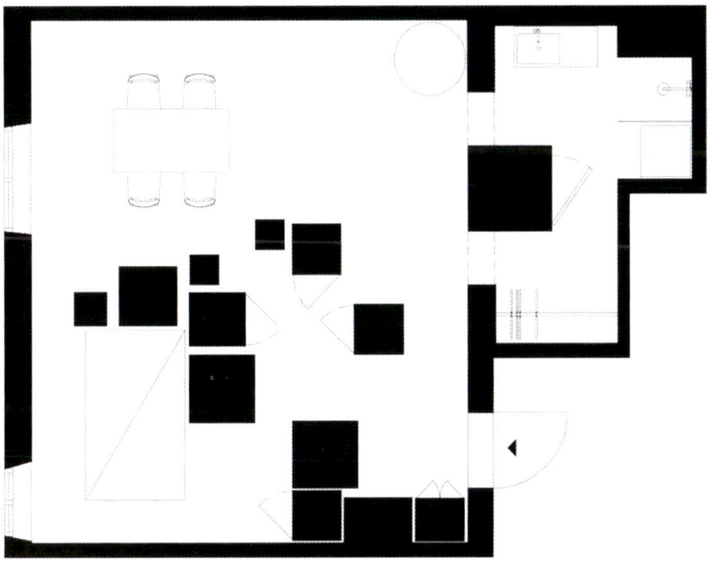

极大的反差往往能促成难以想象的和谐，黑色与白色酝酿的经典数不胜数，这间房子把这番经典再一次推上了极致。金属洁具的简洁造型、考究的抛光，摒弃了繁复的堆砌，冰冷中透出的唯有纯净之景与现代之感。黑色的整体橱柜不时在合页出展露白色的线条，极具装饰感，同时又不失整体的严谨。

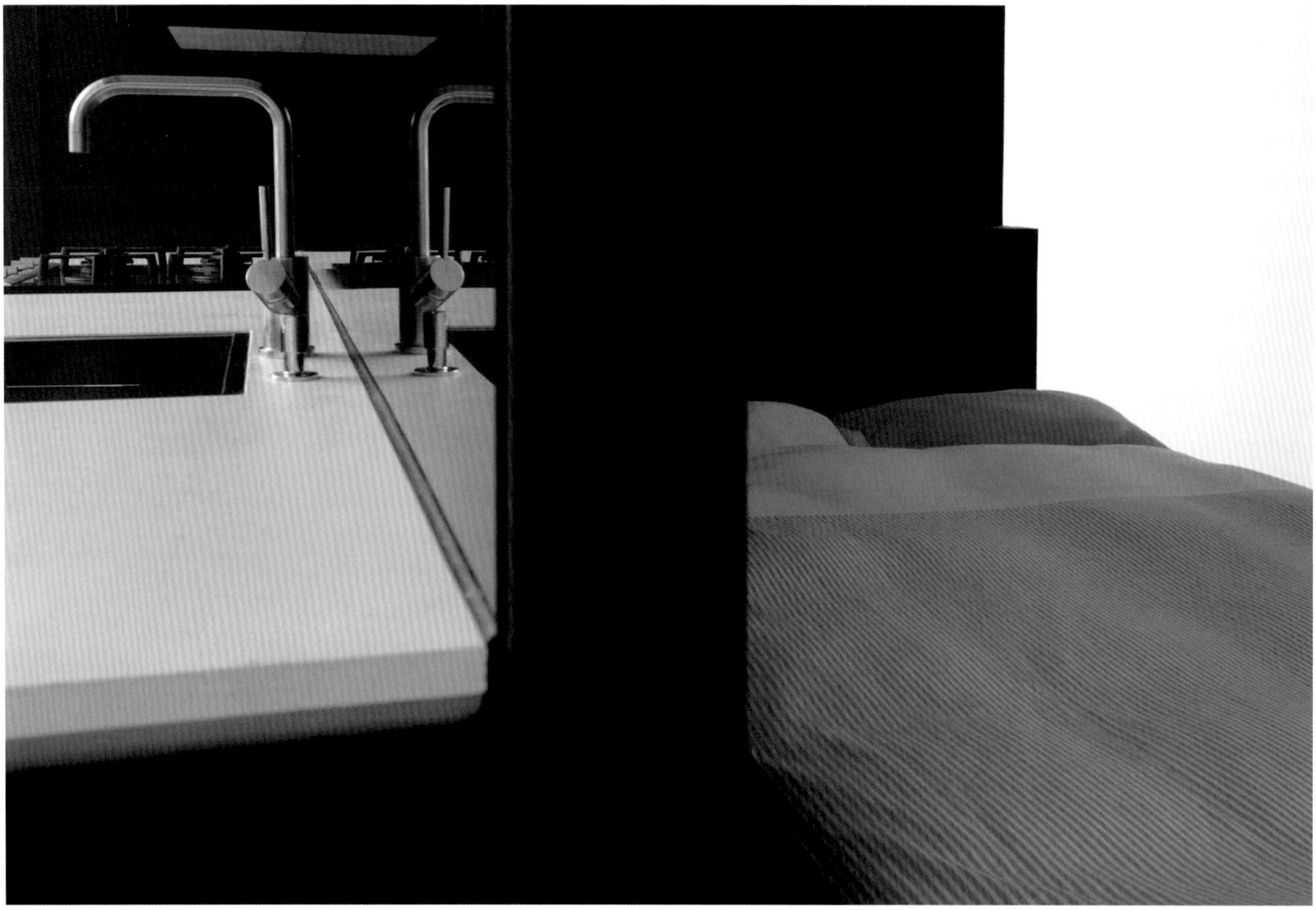

Project: Dupli.Casa
Architects: Juergen Mayer H., Georg Schmidthals, Thorsten Blatter, Simon Takasaki, Andre Santer, Sebastian Finckh
Location: Private House, near Ludwigsburg, Germany
Building area: 569 sqm
Photographer: David Franck

The geometry of the building is based on the footprint of the house that previously was located on the site. Originally built in 1984 and with many extensions and modifications since then, the new building echoes the family archaeology " by duplication and rotation. Lifted up, it creates a semi-public space on ground level between two layers of discretion. The skin of the villa performs a sophisticated connection between inside and outside and offers spectacular views onto the old town of Marbach and the German national literature archive on the other side of the Neckar valley.

Dupli.Casa

项目名称: **Dupli.Casa**
建筑设计: **Juergen Mayer H., Georg Schmidthals, Thorsten Blatter, Simon Takasaki, Andre Santer, Sebastian Finckh**
项目地址**: Private House, near Ludwigsburg, Germany**
建筑面积: **569** 平方米
图片: **David Franck**

建筑在几何学的基础上，以其独立个体的运动足迹重复。原建于1984年，有许多扩展内容自那时以来就开始修改，新的建筑与之相呼应。设计方案通过重复和转并进行了抬升，它创建了一个地面之间的随意层分布的半公共空间。别墅的外装饰上突出表现的是马尔巴赫老城区和德国对面的内卡河谷内外的壮丽景色，提供先进的联系与互动。V

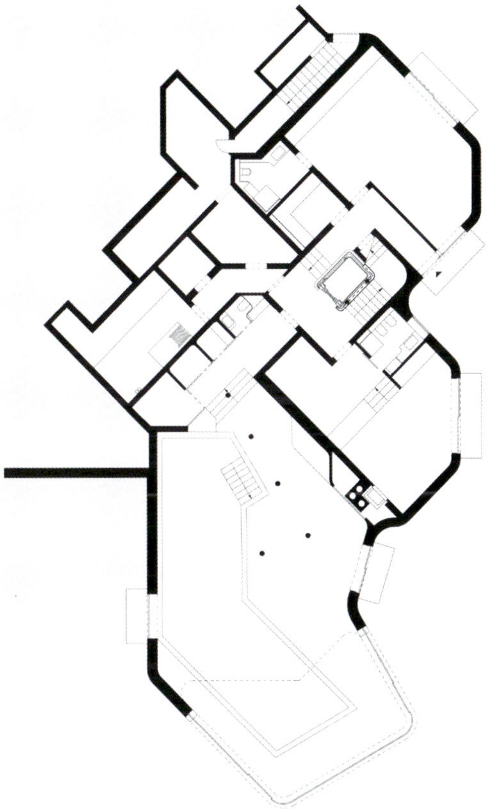

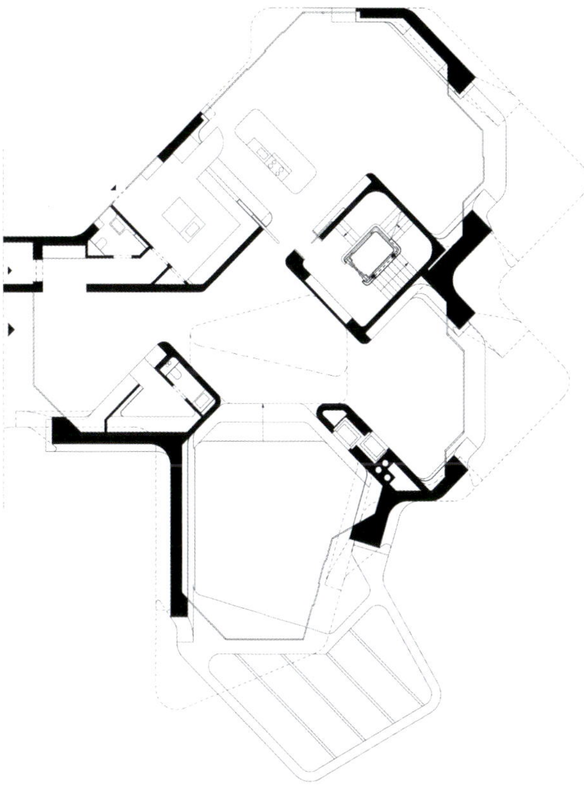

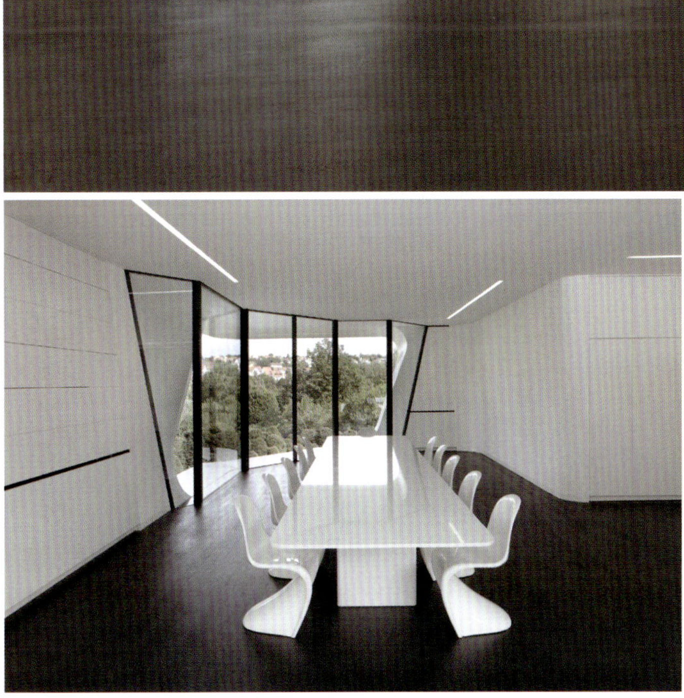

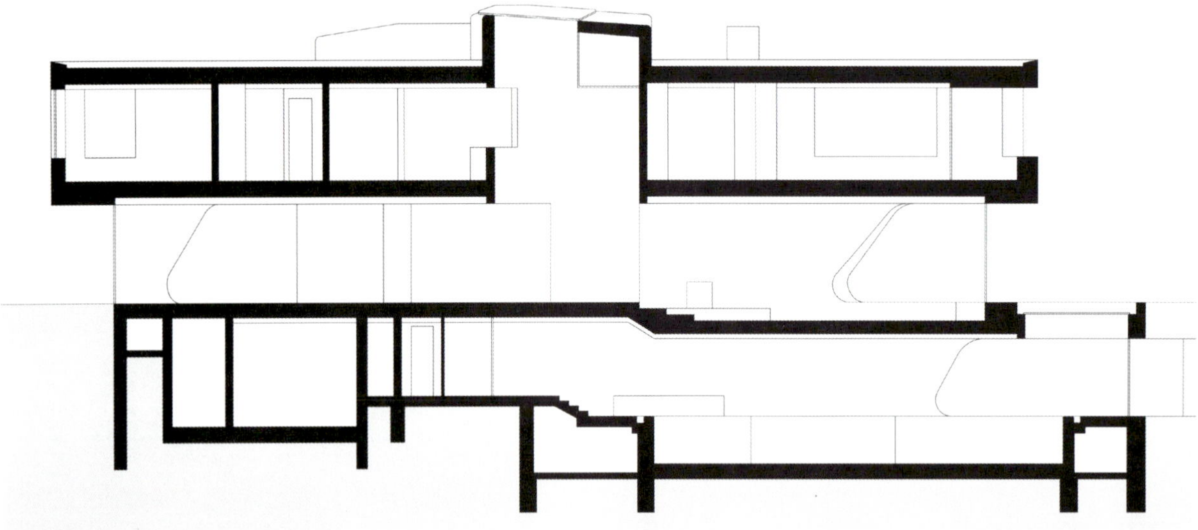

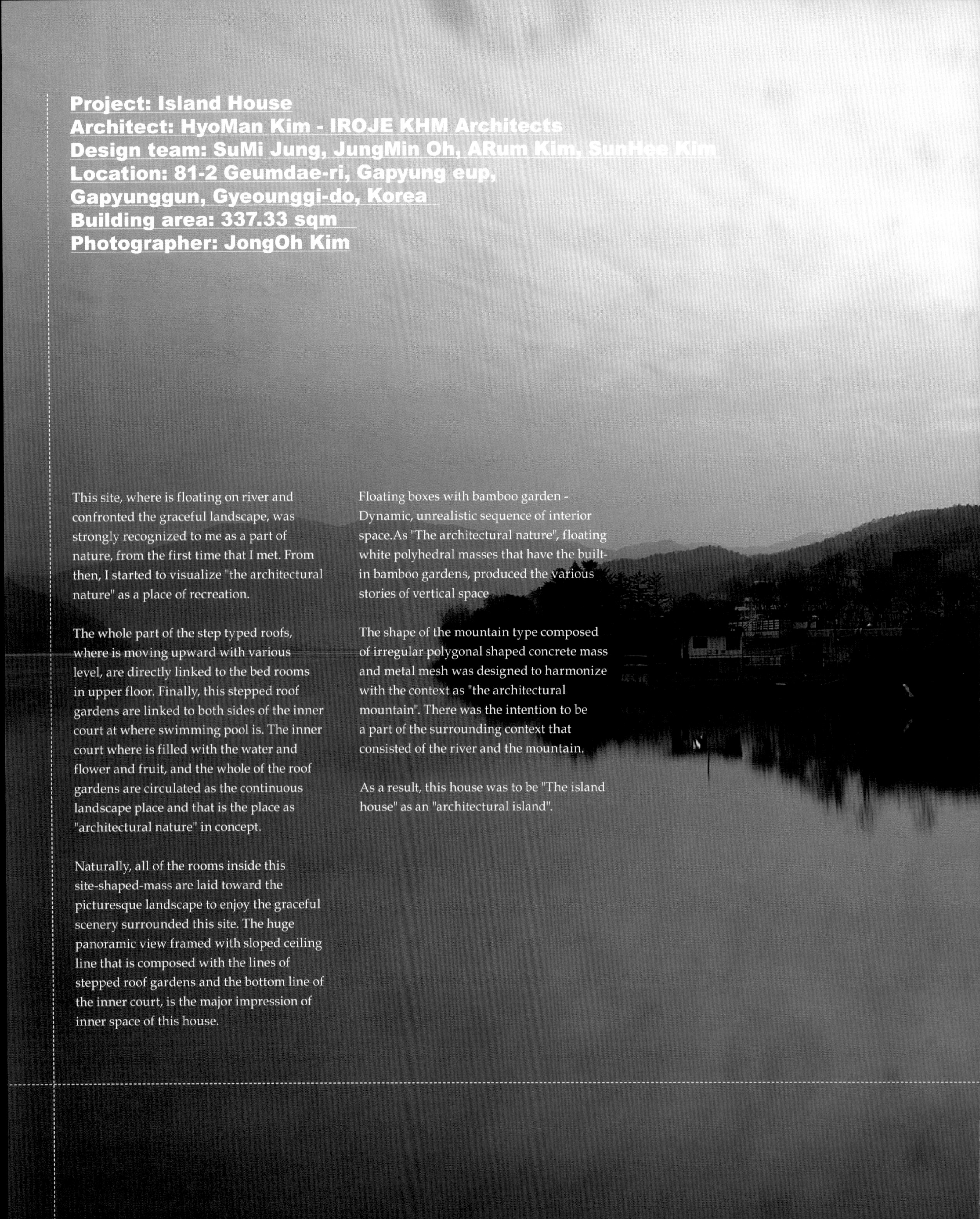

Project: Island House
Architect: HyoMan Kim - IROJE KHM Architects
Design team: SuMi Jung, JungMin Oh, ARum Kim, SunHee Kim
Location: 81-2 Geumdae-ri, Gapyung eup, Gapyunggun, Gyeounggi-do, Korea
Building area: 337.33 sqm
Photographer: JongOh Kim

This site, where is floating on river and confronted the graceful landscape, was strongly recognized to me as a part of nature, from the first time that I met. From then, I started to visualize "the architectural nature" as a place of recreation.

The whole part of the step typed roofs, where is moving upward with various level, are directly linked to the bed rooms in upper floor. Finally, this stepped roof gardens are linked to both sides of the inner court at where swimming pool is. The inner court where is filled with the water and flower and fruit, and the whole of the roof gardens are circulated as the continuous landscape place and that is the place as "architectural nature" in concept.

Naturally, all of the rooms inside this site-shaped-mass are laid toward the picturesque landscape to enjoy the graceful scenery surrounded this site. The huge panoramic view framed with sloped ceiling line that is composed with the lines of stepped roof gardens and the bottom line of the inner court, is the major impression of inner space of this house.

Floating boxes with bamboo garden - Dynamic, unrealistic sequence of interior space.As "The architectural nature", floating white polyhedral masses that have the built-in bamboo gardens, produced the various stories of vertical space

The shape of the mountain type composed of irregular polygonal shaped concrete mass and metal mesh was designed to harmonize with the context as "the architectural mountain". There was the intention to be a part of the surrounding context that consisted of the river and the mountain.

As a result, this house was to be "The island house" as an "architectural island".

Island House

项目名称: **Island House**
建筑设计: **HyoMan Kim - IROJE KHM Architects**
设计团队: **SuMi Jung, JungMin Oh, ARum Kim, SunHee Kim**
项目地址: **81-2 Geumdae-ri, Gapyung eup, Gapyunggun, Gyeounggi-do,** 韩国
建筑面积: **337.33** 平方米
图片: **JongOh Kim**

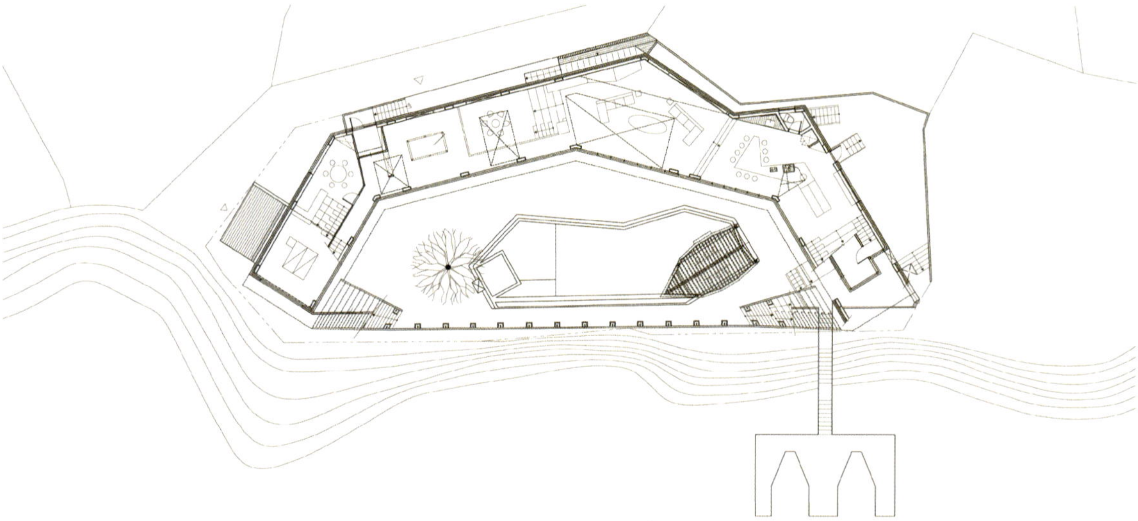

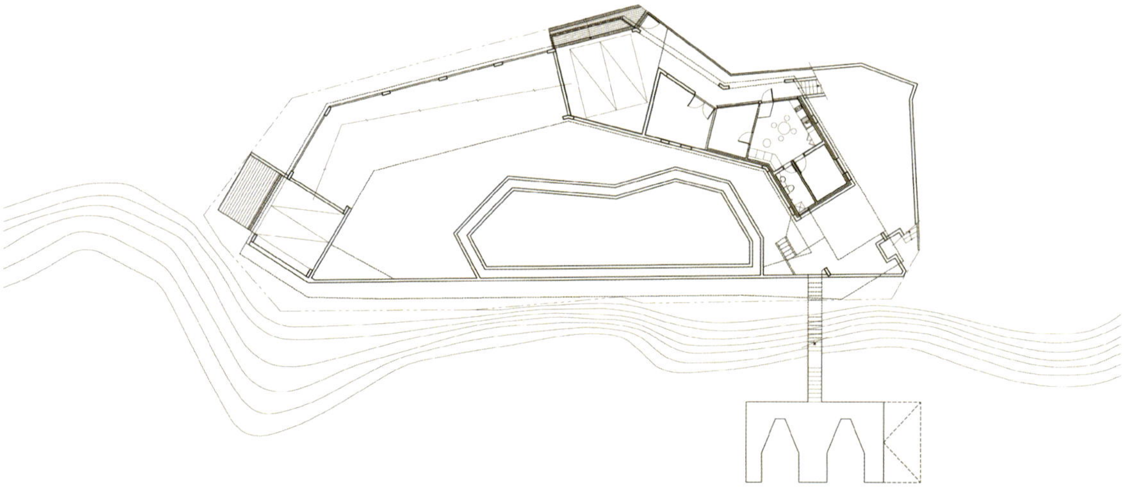

这个网状建筑，像是浮在河面上的优美景观，从我第一次看到就强烈地认为那是一个大自然的一部分。从那时起，我便开始想象这座“自然的建筑”，必会是一个有趣的地方。

该类型的屋顶沿着水平方向上移动，整体的一部分直接连到楼上房间。其中，这是加强空中花园与内院的联系，在这里与游泳池汇聚。其中内院是花卉、水果的空间，以及充满整个屋顶花园的连续景观，那就是作为“自然建筑”概念的地方。

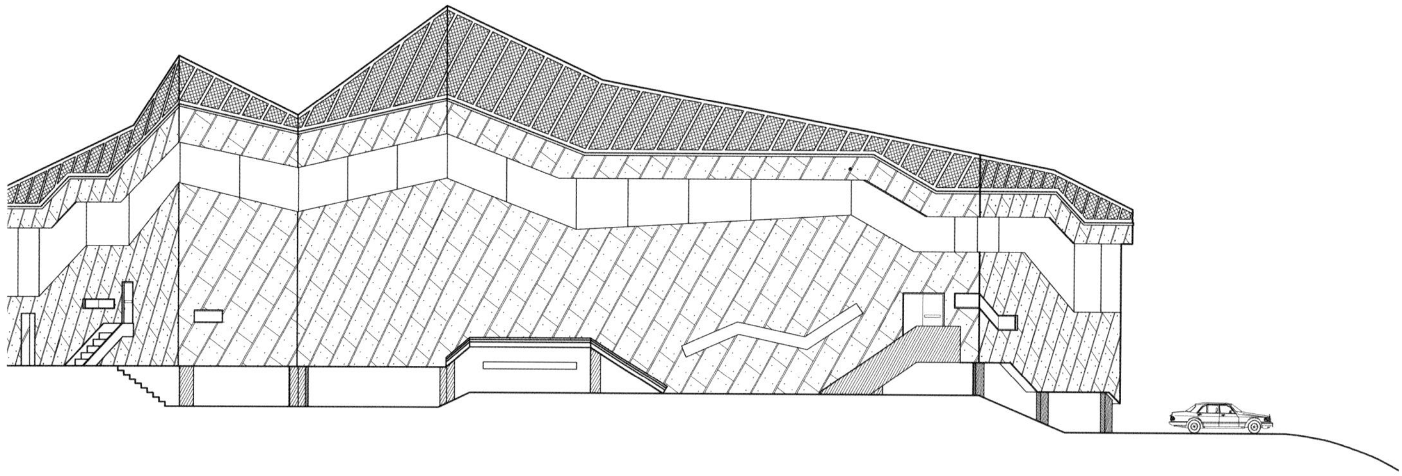

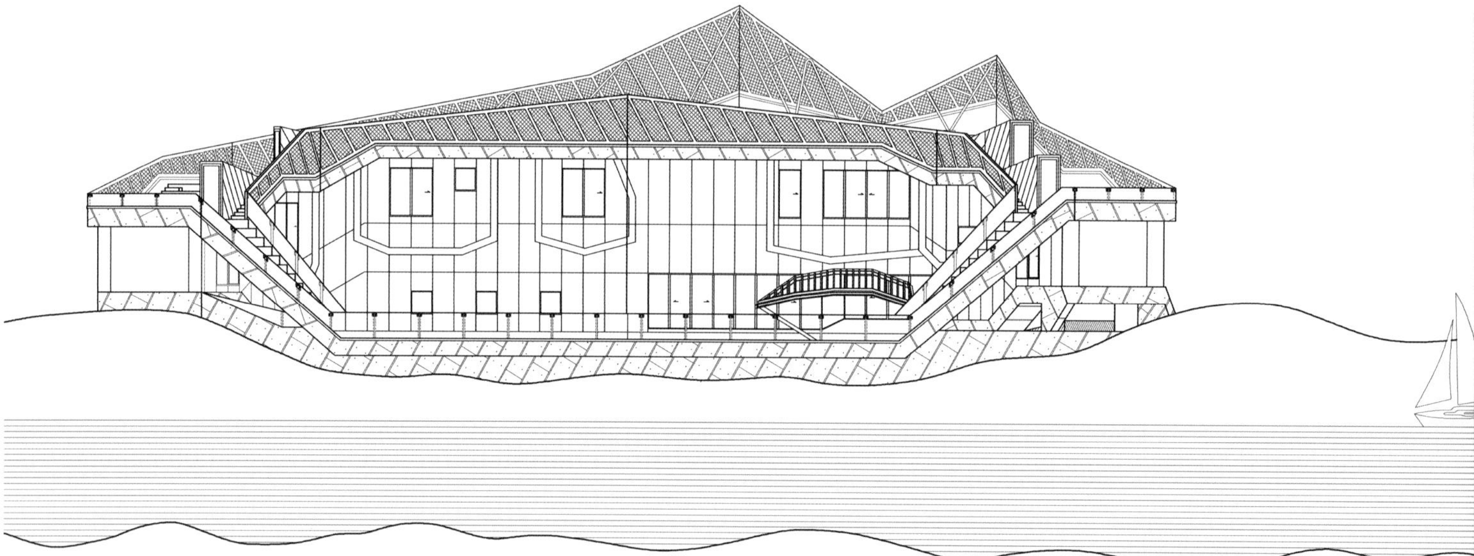

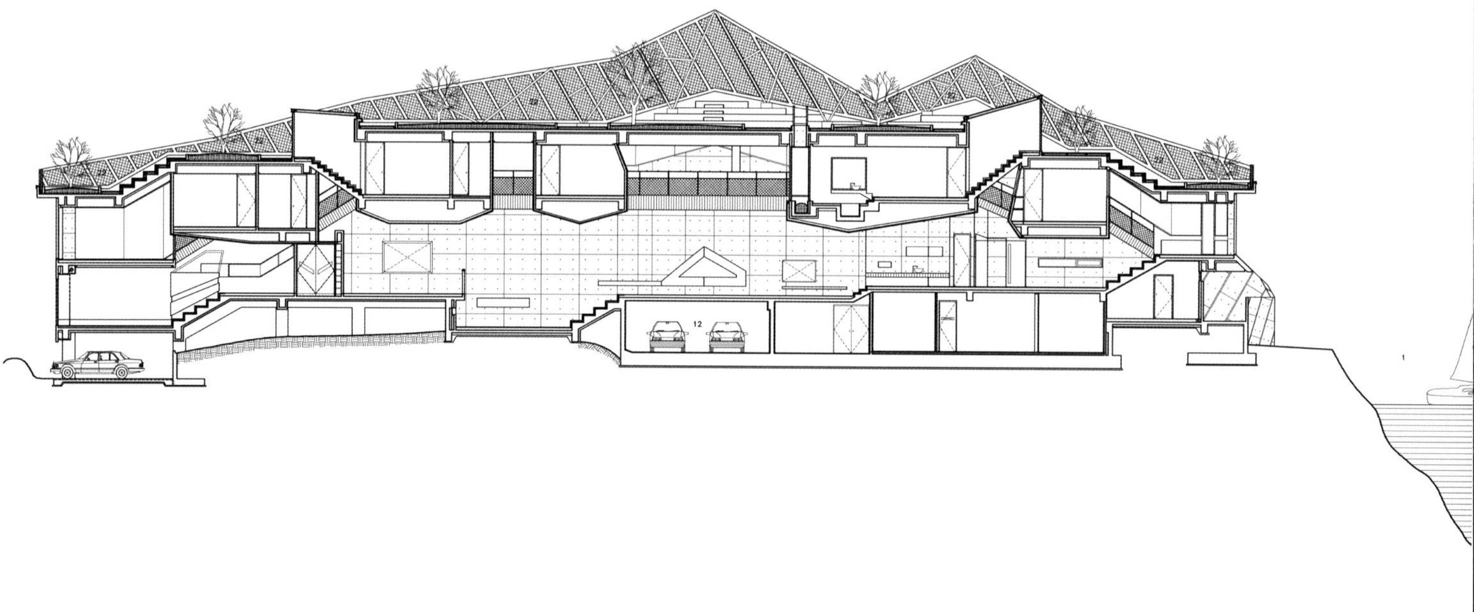

当然，这个网状建筑，内部的房间都面向美丽的风景，优美的风景包围着这个网状建筑。巨大的全景倾斜天花板与阶梯屋顶花园和内院底部组成的框架，形成了这所房子的内部空间的印象。

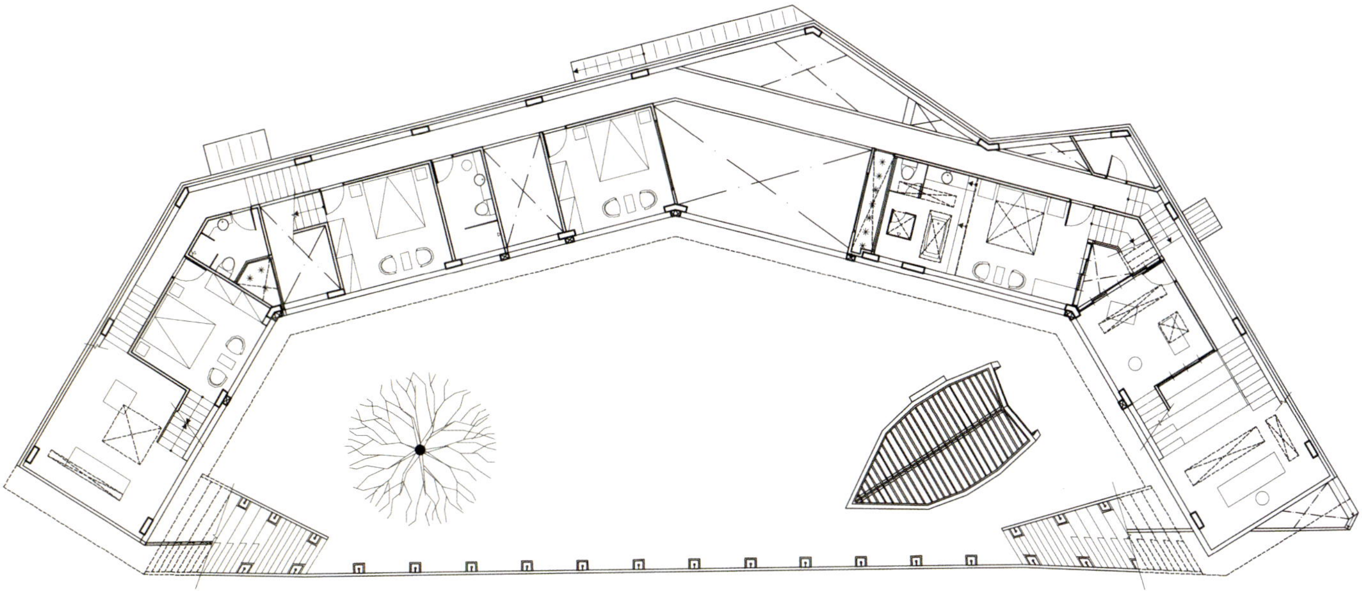

竹制花园的浮动安置，让内部的建筑突显自然，它是由浮动的白色的多面体组成，脱离规则的顺序种植的竹子园林，产生了垂直空间的各种延伸。多边形和不规则形体混凝土和金属网组成山的形状，目的是协调与“建筑之山”的立意。有意向成为周围映衬我们的河流和山脉的组成部分。
因此，这间屋子称为“海岛之家”或“建筑之岛”。V

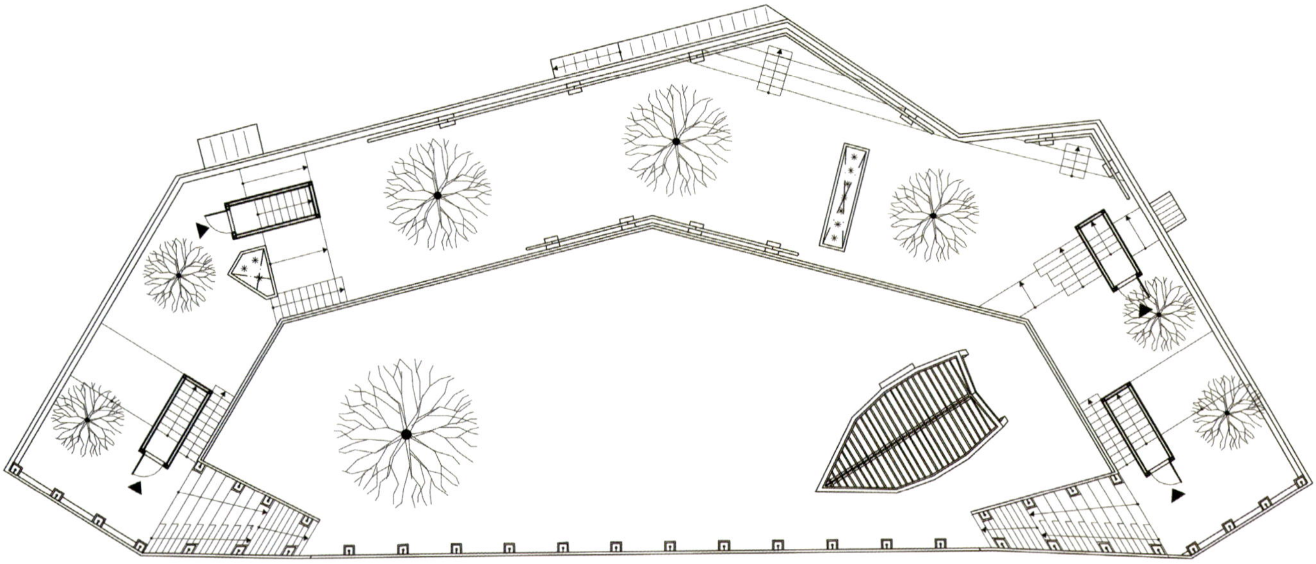

Project: Purple Hill House
Architect : HyoMan Kim - IROJE KHM Architects
Design team : KyeongJin Jung, JiWon Ym, EunHae Park
Structure designer : MOA. SungYeong Oh
Location : 765-15 Sinbong-dong, Suji-gu, Youngin, Gyeounggi-do, Korea
Building area : 110.28 sqm
Photographer : JongOh Kim

This scenic site is a part of the residential area developed in the natural greenery area of the mountain gwanggyo and that is located at the entrance area to trail to mountain.

From agony about the direction to which major living space see, the plan of this house started. After all, the direction to the opposite mountain that will be preserved in the natural green space forever was selected. But this charming selection that introduce the permanent natural landscape brought the problem, bad solar condition, that the major living space is faced to northwest sunlight. And so, we must make the solution to absorb the southern and eastern sunlight into the inside of the house.

"Glass box of light" that will absorb the useful sunlight was introduced to inside space of the various level and these boxes which are planted with flower and fruit are functioned as "Floating glass garden". Actually, we solved the problem and we own both the two conditions that are the introduction of landscape and ecological environment and naturally, we have been reached to the complex solution that all the rooms have their own garden.

The dynamic, unrealistic interior space which are produced with several floating glass boxes over the major living space that is opened to the vertical height of 3 floors, will be the symbolic impression of this house.

We expect that this landscape architecture will coexist with the surrounding context for a long time, as if a small purple hill....

PU

ple Hill House

项目名称: **Purple Hill House**
建筑设计: **HyoMan Kim - IROJE KHM Architects**
设计团队: **KyeongJin Jung, JiWon Ym, EunHae Park**
结构设计师: **MOA. SungYeong Oh**
项目地址: **765-15 Sinbong-dong, Suji-gu, Youngin, Gyeounggi-do, Korea**
建筑面积: **110.28** 平方米
图片: **JongOh Kim**

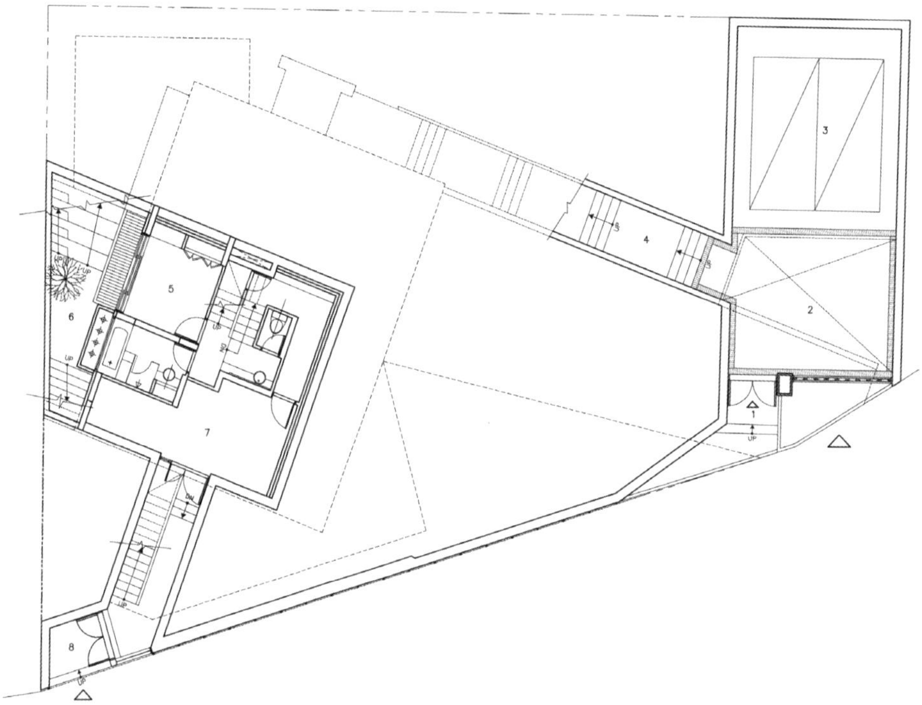

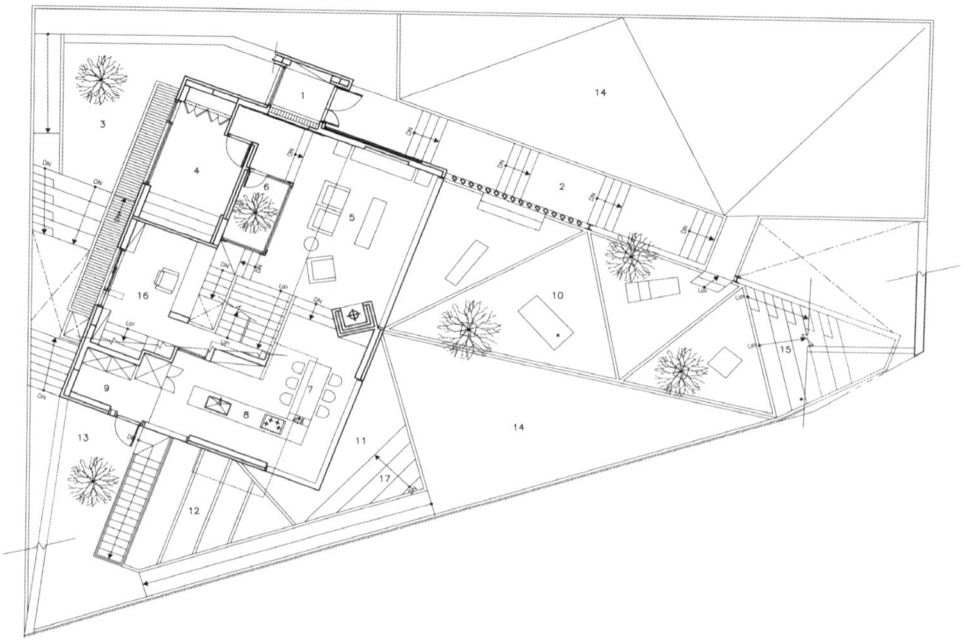

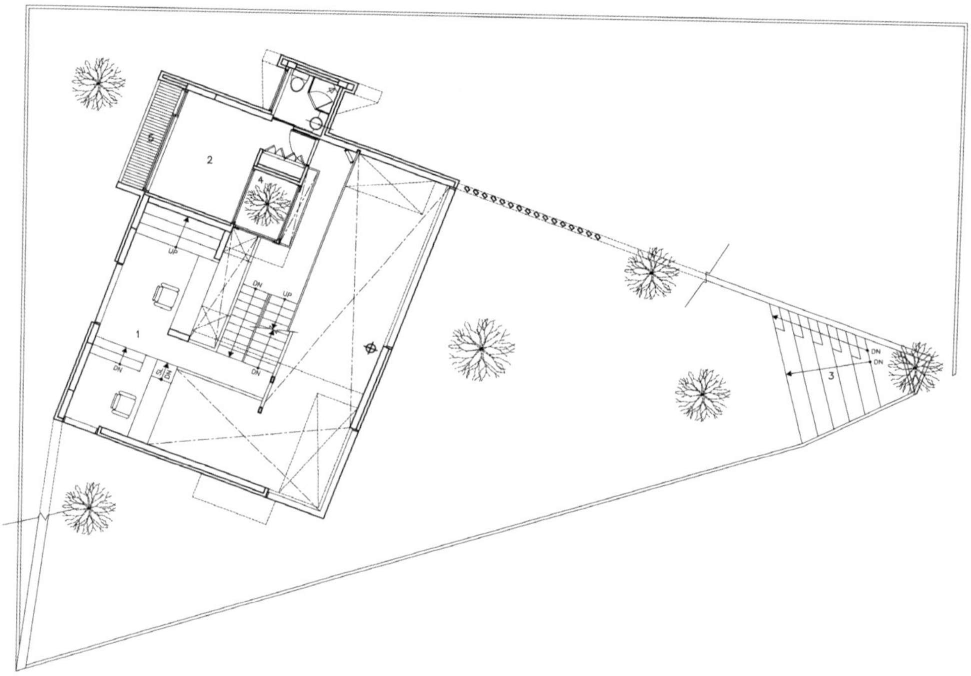

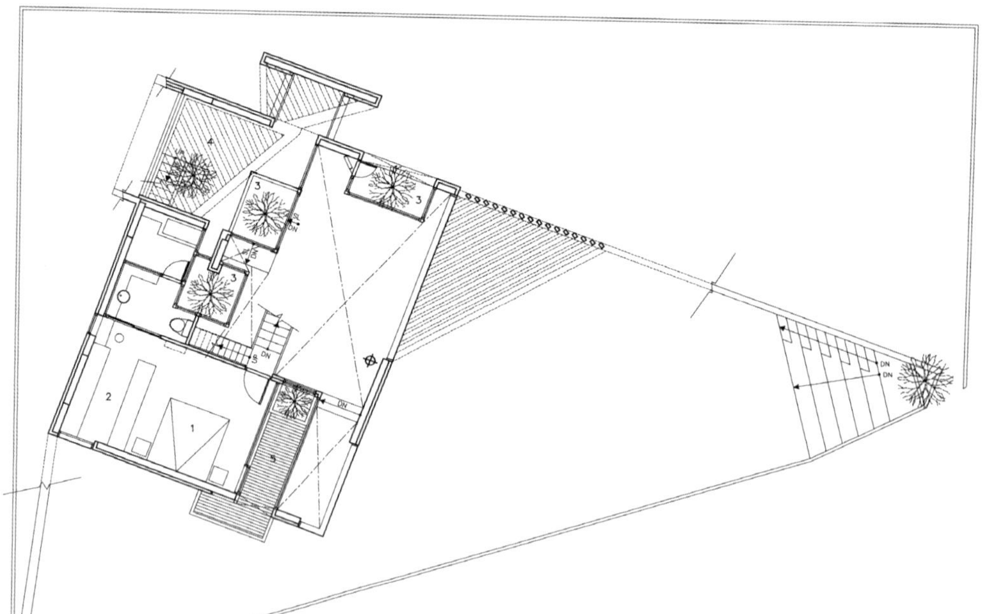

这是一个旅游景点，在自然的绿地面积开发的住宅小区的一部分。房子属于步道入口区，位于山区之内。

这所房子主要的生活空间都是开放式的。毕竟，对面的山永远是自然的绿色。但这种选择，对介绍迷人的自然景观带来很大的问题。主要的生活空间是面向西北的，太阳无法直接照射。因此，我们必须解决采光的问题，将南部和东部的阳光吸收到房子里面。

“充满光线的盒子”中，能吸收的太阳光，分给内部空间的各个层面，这些种植花卉和水果的箱体运作过程被称为“浮动的玻璃花园”。

其实，我们解决了这个问题，我们同时拥有这两个条件，即：景观和生态环境、自然的缔造，我们已经采用了复杂的解决方案，所有的房间都有自己的花园。

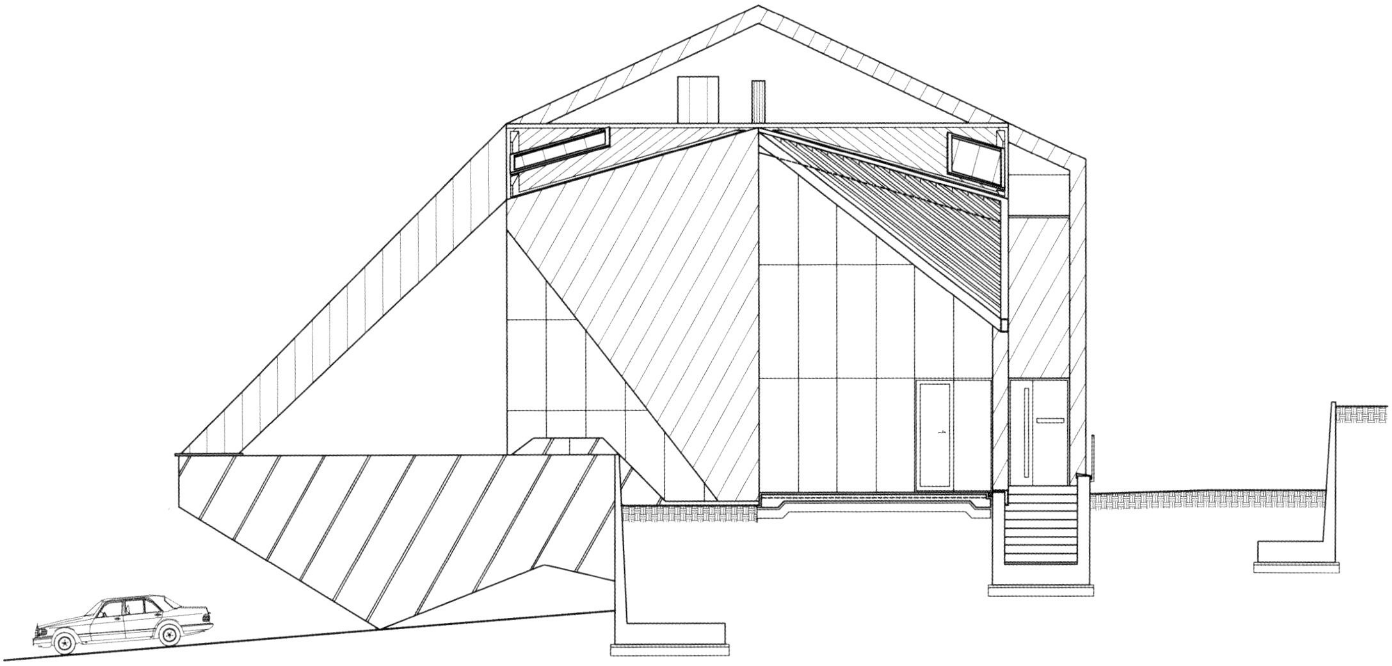

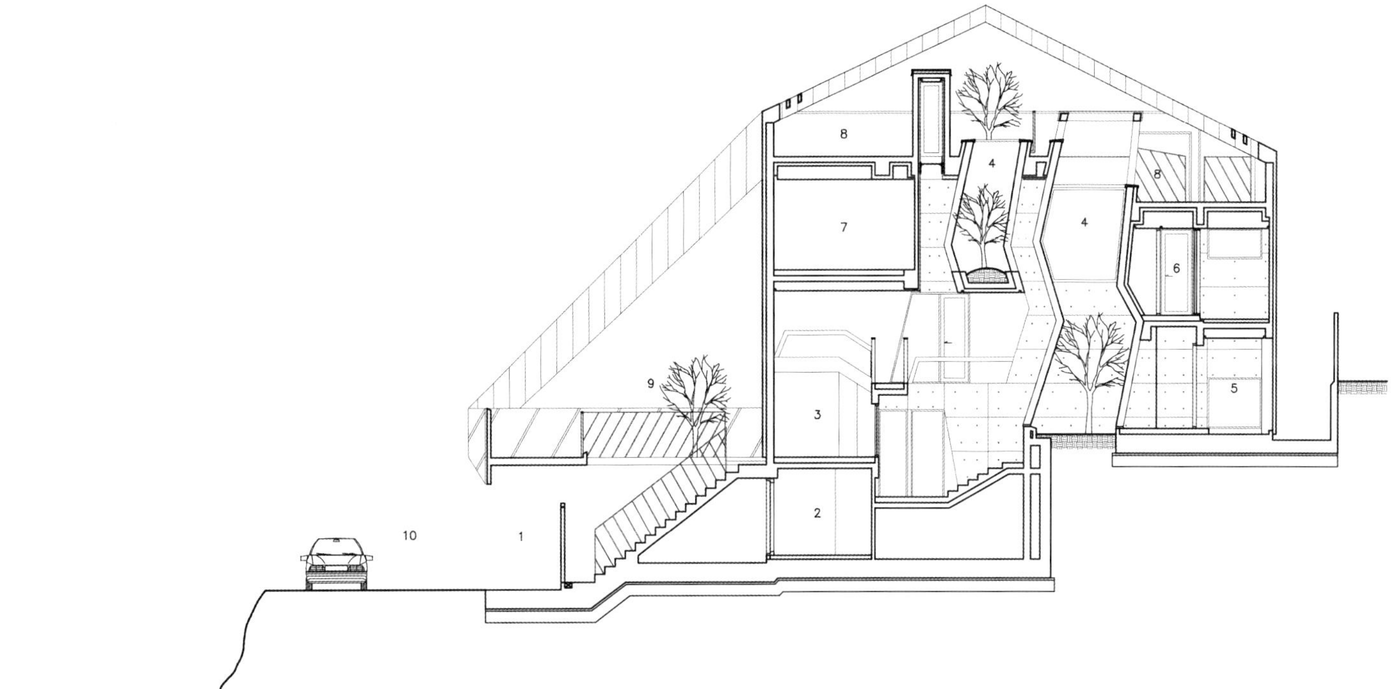
8
4
8
7
4
6
3
9
5
2
10
1

浮动的竹子花园盒子——动态和垂直空间序列，住宅贯穿3层的主要生活空间中的几个浮法玻璃盒子产生了动态，那是充满幻想的室内空间，这成为住宅的象征印象。我们希望景观设计与周围的环境共存很长一段时间，仿佛一个自然形成的小紫山....V

Project: Casa Paracaima
Architects: DCPP design firm
Location: Mexico City, Mexico
Designer: Pablo Perez Palactos, Alfonso De La Concha Rojas
Photographer: Luis Gordoa Hernandez

Casa Paracaima is located in Mexico City in a private residence, designed by local design firm DCPP, completed in 2008. In a house can be from every point of view presents a different face, can be called a house of 360° around the type. Problems of modern architecture is not to re-arrange its lines, not to find a new decoration, new door frames, or else a kind of new form, but to learn science and technology, a great achievement for each, the structure of the new buildings to a reasonable level, the establishment of a new form, new lines and new raison d'etre. This depends entirely on the special conditions of modern life, and it extends to our sense of aesthetic value.

Casa Paracaima consists mainly of two spaces, although there is an area of 400 m^2, but its sides and in front of a lot of restrictions. But designers are not bound by these limitations, but local use of these constraints to design, continually seek to maximize the housing space and the ideal towards. Architecture is to create space for artistic space, rather than outlining the facade. The outer space defined by the wall, while the space is through the complexity of the wall shown. The building has a large, structural arch, designers use it to carry more space. The external region of the main building has an open side of the 7.5 m high to connect upper and lower levels of space, while the other side of the library to use. The two main indoor space for people is the impression of two distinct styles.

In this building, the space is the first, the wall as the creator of the importance of form and proportion of supremacy in all the above systems, concrete walls are not exposed at all times to express his own bright and clean with the United States. Casa Paracaima designers use design language simple, concise, in addition to the functionality-oriented architecture, language variety of materials and excellent workmanship makes up for this lack of modern building decoration, and their sense of the rich far more than decorative.

sa paracaima

项目名称：**Casa Paracaima**

建筑设计：**DCPP**设计公司

项目地址：墨西哥墨西哥城

设计师：**Pablo Perez Palactos，Alfonso De La Concha Rojas**

图片：**Luis Gordoa Hernandez**

Casa Paracaima是坐落在墨西哥城的一所私人住宅，主要包含了两个空间，虽然占地面积有400m^2，但是它在两侧和正前方有很多限制。然而设计师并没有被这些局限拘束，反而就地利用这些限制来设计，不断地寻求房屋空间的最大化和理想朝向。建筑的艺术手法是创造空间而不是勾画空间的立面。空间的外围是由墙确定的，而空间也是通过该墙的复杂性表现出来的。该建筑拥有一个巨大的、结构性的拱形，设计师利用它来承载更多的空间。建筑的主要对外区域的一边有一个开放性的7.5m高连接上下两层的空间，而另一边则作为图书馆使用。室内的两个主要空间给人们的也是两个截然不同的风格印象。

在这栋建筑中，空间是第一位的，墙作为形式创造者的重要性以及系统比例凌驾在一切之上，混凝土的墙面无时无刻不在裸露地表达着自己的光洁美。Casa Paracaima的设计师运用的设计语言简洁、洗练，除了注重建筑的功能性，语言丰富多样的材料和优异的做工也弥补这座现代建筑装饰的不足，并且它们在丰富感上远远超过了装饰。

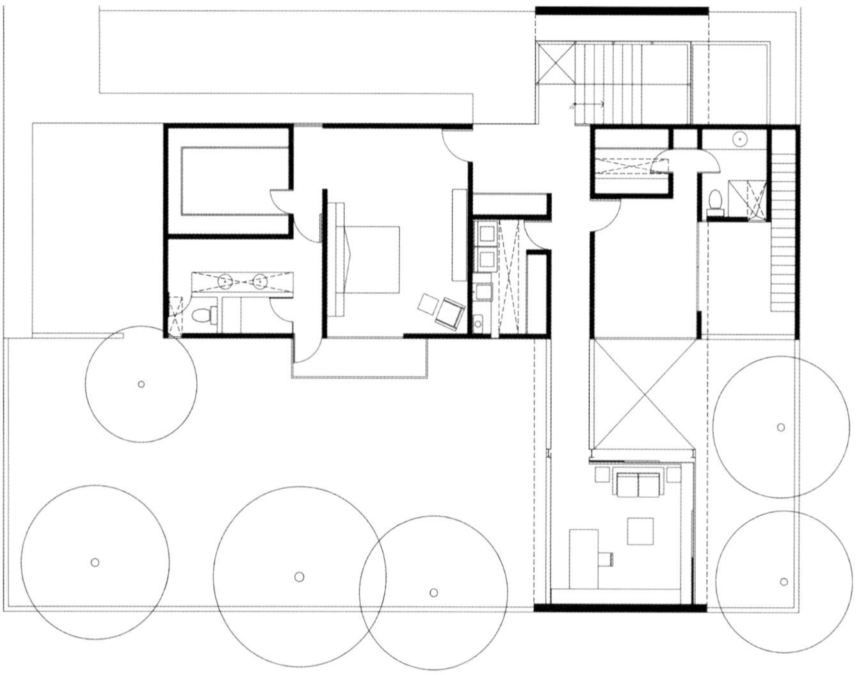

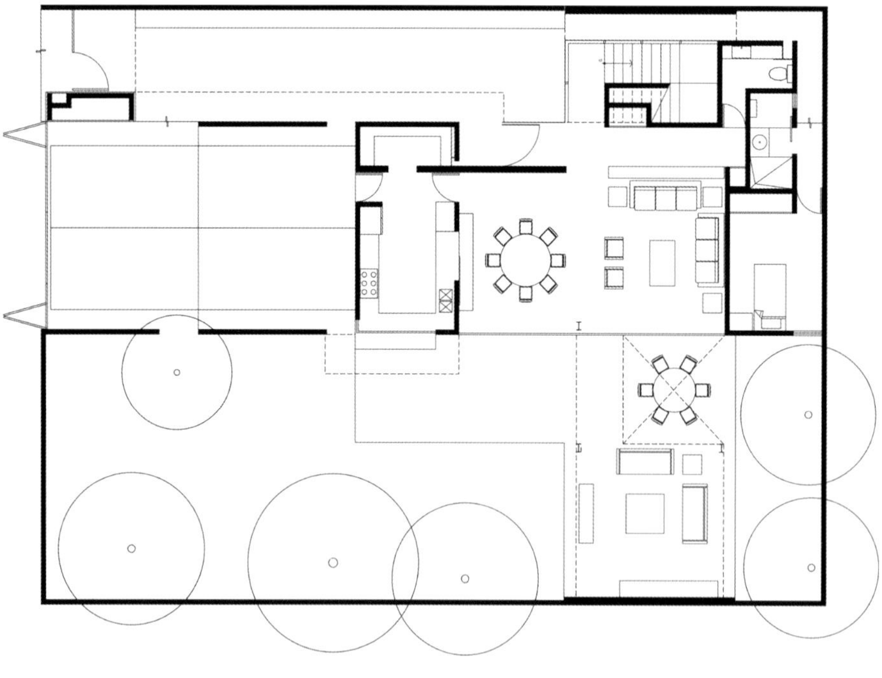

在建筑中，有两种做到忠实的必要途径。一种必须忠实于建设纲领；二是必须忠实于建造方法。忠于纲领，指的是必须精确和简单地满足需要提出的条件。忠实于建造方法，指的是必须按照材料的质量和性能去运用它们，对称性和外观的形式等纯艺术的问题在这些主要的原则面前只是次要的。V

Project: L-Shaped House
Architects: Titus Bernhard Architekten BDA
Location: Germany
Building area: 741 sqm
Photographer: Christian Richters

For the home owner is looking for: multi-storey villa, with approximately 700 m^2 of floor area, can be a distance, about three kilometers from the Danube scenery. But for us, this land of high-quality stems from its unique terrain and openness, but not because of the Danube where you can enjoy the scenery, not to mention views of the Danube is more than one lane of traffic routes and the city Some small and medium sized building block, one can only watch this part of the landscape but not all, of its landscape.

Response options for this situation is a paradox program, which not only take into account the natural landscape, but also to provide changing according to different themes inward-looking courtyard. We are a design as those based on the unique terrain, diversity and complexity of the construction environment made just the right Xianyou building form, building types Jinxingtiaozheng seem not Namegeli: this way, House L Jiang Cheng Wei's a strange, Xiang Mu in generated and self-contained contradictions structure type. Owners will receive a kind of similar to the "Matrix" or the square format of the spatial structure, in such a structure, they can be personalized to change and adjustment, and that change and adjustment at the same time both a higher value and diverse atmosphere.

Response options for this situation is a paradox program, which not only take into account the natural landscape, but also to provide changing according to different themes inward-looking courtyard. We are a design as those based on the unique terrain, diversity and complexity of the construction environment made just the right Xianyou building form, building types seem not Namegeli: this way, House L Jiang Cheng Wei's a strange, Xiang Mu in generated and self-contained contradictions structure type. Owners will receive a kind of similar to the "Matrix" or the square format of the spatial structure, in such a structure, they can be personalized to change and adjustment, and that change and adjustment at the same time both a higher value and diverse atmosphere.

haped House

项目名称: **L-Shaped House**
建筑设计: **Titus Bernhard Architekten BDA**
项目地址: 德国
建筑面积: **741** 平方米
图片: **Christian Richters**

对于该住宅，业主期待的是：多层的别墅，大约700 m^2的使用面积，可以远观约三公里外的多瑙河景色。但对我们而言，这一地块的高质量源于其得天独厚的地形及开阔度，而并非因为在此可以观赏多瑙河的景色，更何况多瑙河的景色被一条多车道的市内交通要道和一些中小型建筑物遮挡，人们在此只能观赏到其部分景观而非全部景观。对于这一情况的应对方案是一个矛盾方案，它不仅要考虑到自然景观，而且要提供可以根据不同主题变化的内向型庭院。我们的设计理念基于该地所特有的地形，多样而复杂的施工环境使得单纯的对现有建筑形式、建筑类型进行调整显得不那么合理：如此一来，House L将成为一个奇特的、项目中产生的、并且自成一体的矛盾结构类型。业主将会获得一种类似于“矩阵”或者说是方格式的空间结构，在这样的结构中，他们可以进行个性化的改变与调整，而这种改变和调整同时兼具较高的使用价值和多样的氛围。

该项目场地被分割为几处单体建筑区域，这些单体的建筑部分在这个清晰界定的场地内部被整合为一个整体。不断变化的景观、内外视角和错层的设计营造了空间多样性，使得空间的流畅与连续成为可能。然而，该项目区域仍然总体呈现出封闭、完整、自成一体的格局。每一个单体建筑区域都拥有一处内向型的天井，天井成为一个高质量的、私密的而又不受天气影响的休息区域。大范围活动的入口区域使得私人空间可以更加灵活地变换，尤其是可以根据阳光照射的高度调整遮阳装置。由此，局部的景观、光线以及阴影共同产生了极其吸引人的效果。坡道部分的形态以及面向坡道的开放庭院，更能使人感受到该住宅所处的独特地形，这样“私密性”与“开放性”之间的变化也更加灵活。V

Project: Lim Geo Dang
Location : JangHangDong, IlSan, GoYang, Korea
Building area : 107.77 sqm
Photographer : YongKwan Kim

The spatial design concept set for this small house was based on is dramatic circulation, which was characteristic of traditional architecture in Korea. The circulation of ancient Korea continues infinitely in space, overlaps and penetrates each other asymmetrically, and finally, produces several picturesque frames by surrounding nature and building.

At this time, the house has to be a shelter from the outer world. It is necessary to contact with nature which brings the urban house rich changes and variation during its life time(i.e., witnessing the blossoming and decaying of flowers). "Madang"(open court yard) was the very "nature" in ancient Korea.

Eight places of "madang" located in different levels were planned to produce attractive and interesting drama of spaces. All the major spaces of this house were designed to have a direct relationship to "madang". Entrance court, basement courts Ⅰ&Ⅱ, jungja(pavilion), service court, inner court, and upper court Ⅰ&Ⅱ were all the "madangs" at different levels that linked each other with both horizontal and vertical circulation. There are interesting changes, variety, through the circulation between the "madang".

It is a very small site to have exterior spaces of desirable size by common layout of rooms at ground level. To have as wide an area of "madang" as possible, only a living room and a dining room are placed on he ground level; the others are either below ground or above ground. Moreover, the bed-zone-mass of the upper floor was built with a piloti structure; therefore, the space below can be used as "madang", too. This so called "RU" type program brings the house several merits of architectural condition. First of all, it gives the inner court a comfortable atmosphere because of the surrounding "madangs" which play important roles. Secondly, it creates another "madang", which was a part of the inner court, covered with a roof. Last, without any fence (a solid fence cannot be built in this area according to the regional building code), it separates the house from public road, and, at the same time, it solves the problem of relationship between the "privacy" inside and the neighborhood outside. "Madang" is a buffer zone between the house and he public road. It is also the place to contact with the neighborhood.

A heavy mass of woods flying over the roadside was expected to evoke the "flying" image from traditional architecture were used in this project. "T" shaped metal ties used to fasten wooden members to each other, over hanging doors, and steel rods which hang out the doors are the traditional elements; however, they can be modernized and recall past as the same time.

Traditional culture is the origin of present ideas. The study of this project was only an experiment to read inheritance in modern language.

im Geo Dang

项目名称: **Lim Geo Dang**
项目地址: **JangHangDong**, 韩国
建筑面积: **107.77** 平方米
图片: **YongKwan Kim**

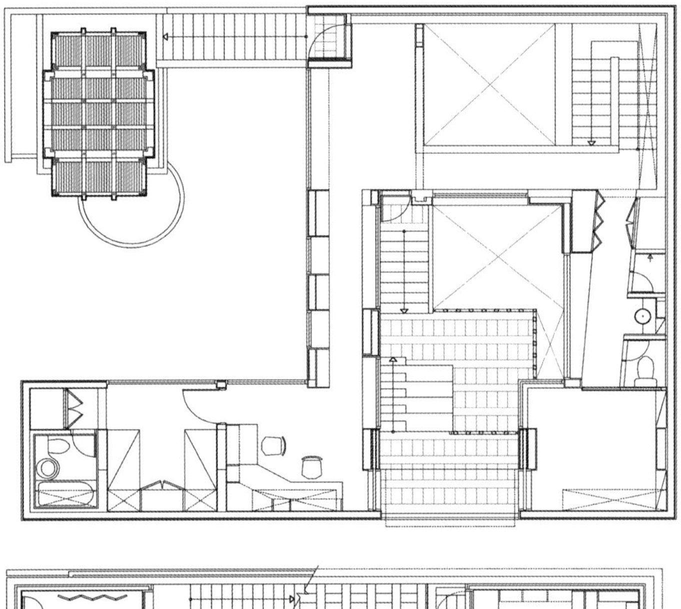

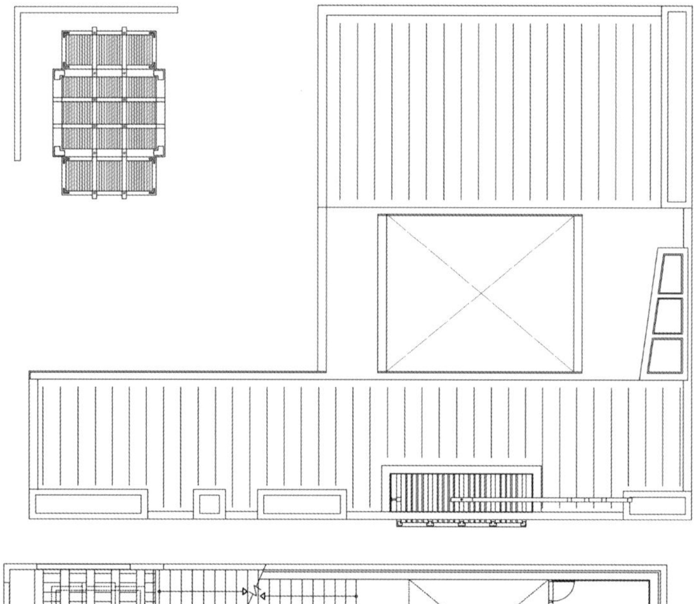

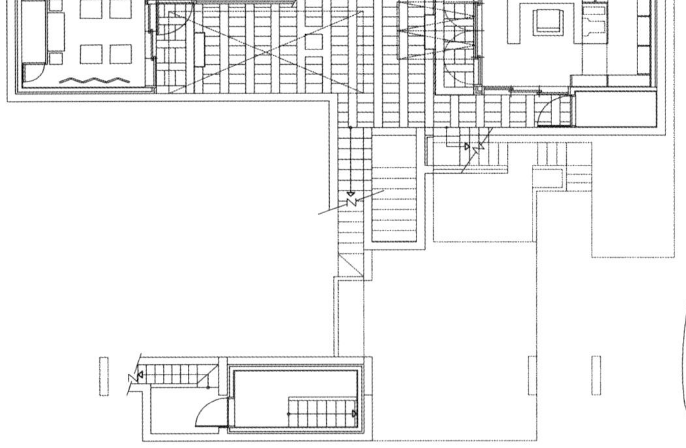

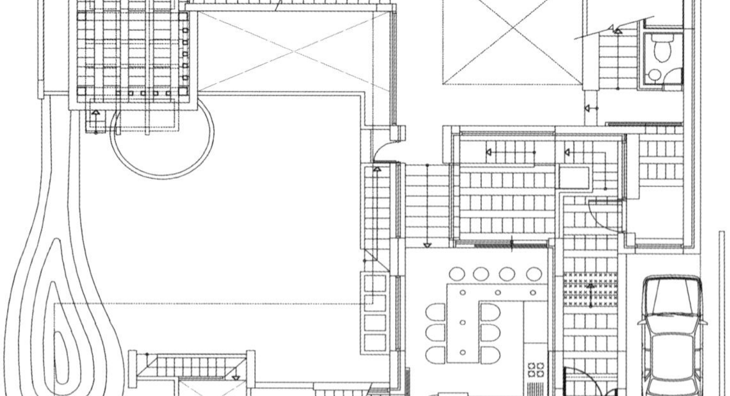

这个房屋的空间设计理念以戏剧的循环为基础，这是韩国传统建筑的特点，古代韩国的循环理念继续在无限的空间内，以不对称的方式重叠渗透，最后由周围的自然环境和建筑产生框景的效果。房屋是外界的避难所，需要在生命周期内参与城市的变迁，“Madang”即：开放的庭院。在古代韩国庭院是非常能表现自然环境的。

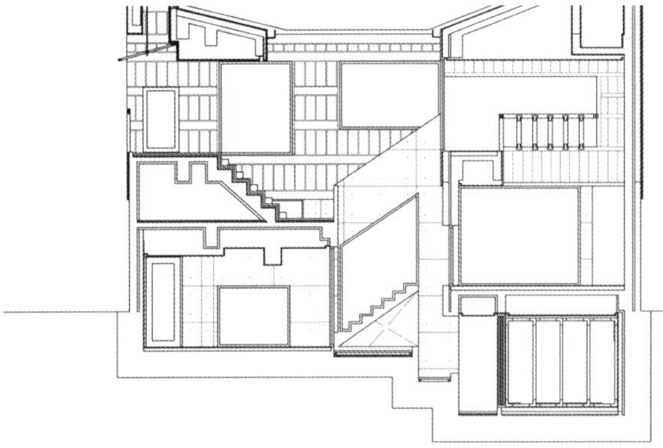

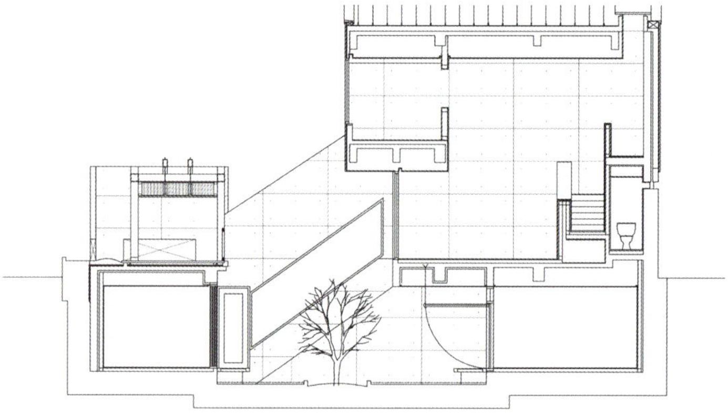

有"Madang"的8处地方位于不同水平线上用作规划，来产生出有吸引力、有趣的戏剧空间，都设计有直接关系的"Madang"。入口、地下室、内院，所有"Madangs"联系在一起，横向和纵向的不同层次的相互流通。还有有趣的变化，通过各种流通之间的"Madang"。

这是一个非常小的空间，有通向地面的理想布局，外部空间。都尽可能广泛地把公共区域作为"Madang"，只有一间起居室和一个餐厅放置在底面，其余的要么在地下，要么在地面以上。此外，卧室的床区，大规模地建有piloti结构，因此，下面的空间可以说是开放的。这种所谓的"儒"型方案将房子的建筑条件等优点强化。首先，它给人的内院围合感，因为周围的"Madangs"发挥了重要的作用，营造出了一个舒适的氛围。其次，它创造的另一个"Madang"，这是内院的一部分，屋顶覆盖其上。最后，没有任何围栏（坚实的篱笆不能在此区域得到施展），它分隔开公共道路的同时，解决了他们之间的"隐私"问题。"Madang"是一个公共缓冲区之间的房子，是一个开放通道，也可以在此与邻居攀谈，用作休闲娱乐。

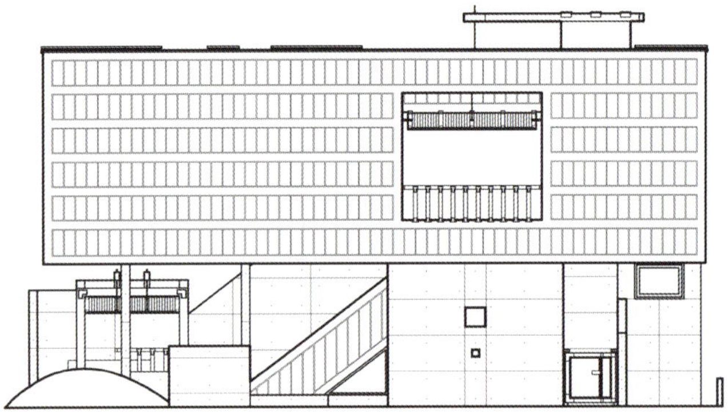

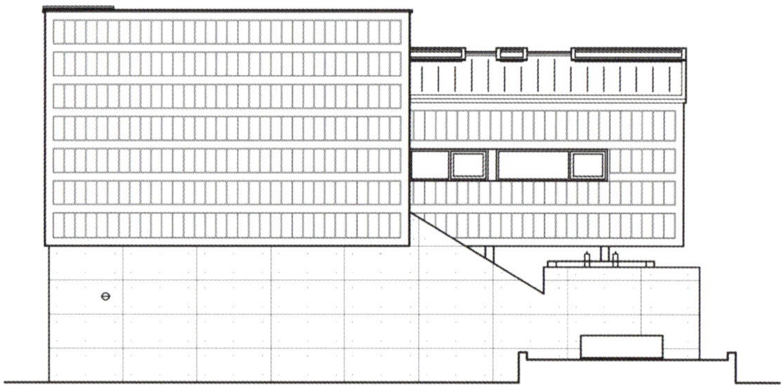

路边的树林里唤起“飞天”的想象，传统建筑的构想在这个项目中得以使用。该项目是一个实验现代传承语言的探究。“T”形金属用于木结构在固定吊门上，挂门是传统的元素，但它们可以被现代化，同时，传统文化是当代思想的起源。V

Project: Jo Rin Hun
Architect : HyoMan Kim - IROJE KHM Architects
Design team : SuMi Jung, MiYeong Park
Structure designer : YeongRok Sin
Location : 15-152, Hyehwa-dong, Jongro-gu, Seoul, Korea
Building area : 169.97 sqm
Photographer : JongOh Kim

The site is located in the outer Bukchon nearby the designated cultural properties of Seoul city like rampart of Seoul. Hyehwamun, Kim Sang Hyeop's House. Many Korean-style houses are existed till now, but remained houses are removed at that with the housing development prevalence of multi-family house last year. So it is one of the villages that are in progress of modernization and Jo Rin Hun is same case that is pressed with the high-storied neighborhood. It was a distressed situation to unavoidably remove the existing Korean traditional house as a position of culture destroyer.

The character of history and place of the Korean-style house inherited spatially by composing the 'garden' of existing house for 'garden' of Jo Rin Hun. This garden functions as a spatial element that satisfies the right of sphere and ownership with the recognition that each of the household living in this house is the owner of each separate house. Many detached house composed vertically with outside-stair as a passage and the small outside space will be a 'city windpipe' that connects that city and architecture strongly.

By incoming the neighbored landscape with the expended metal translucent board, the light, wind, sound through the small vertical courtyard surrounded by each houses are effective and forming introverted calm spatial environment. It is intended to function hiding neighborhood, filtering surrounding landscape, control of the light by me lid of the outer cover adjoins neighborhood with translucent skin. Indistinct landscape of the village, clearly visible shape of the Korean-style house, the whole views of Seoul with Namsan tower and the festive night views were the landscape program of Jo Rin Hun. The city and architecture are endowed with strong mutual response and finally this mass become to carry the un-architectural property of matter of transparence introversively, translucence extrorsely. It is intended to feel Jo Rin Hun, which is vertical and huge mass comparatively, as 'un-architectural' property of matter to harmonize with the horizontal stable landscape formed by the remaining Korean-style houses and to form a new city context that corresponds to the change. It lost by the shaded portion of road, right to enjoy sunshine, cultural property protection. By indoor planting to the remaining mass, It could be recognized as an ecological mass, as well as, the whole could recognized as if translucent/opaque un-architectural object are covered with expended metal and intended to grant a formable sensitivity that harmonized with the image of Korean style house to the structure of skin.

I aimed at nature-friendship by introducing the nature positively and aimed lies latent a small and abundant spatial story by residence together with the large and small houses.

By openly establish the secret private life programs, that is, concealable action apparatus like shower, bathtub, close-stool, washstand, he/she could liberate from restraint of secrecy by him/herself and could present action of life. With the preparation to feel the instinctive pleasure in the nature that is an instinctive environment, all the action of him/herself become a sight and concerned object. So it was an interesting work to plan the 'place' where exist a history and story to aware of the vestige of life always.

Such program is small, but could be said it is large, because there are a lot of contributions that sensed as large space and form a various/dramatic space, reduction of construction expenses.
The construction expense is a mountain to go over. It was effective for cost administration to minimize the expenses with the effort by using cheap materials like concrete floor, wall finishing, dryvit excluded finishing materials, prefabricate sandwich panel and simplification of process.

JO RIN HUN

项目名称: **Jo Rin Hun**
建筑设计: **HyoMan Kim - IROJE KHM Architects**
设计团队: **SuMi Jung, MiYeong Park**
结构设计: **YeongRok Sin**
项目地址: **15-152**, **Hyehwa-dong, Jongro-gu,** 韩国
建筑面积: **169.97** 平方米
图片: **JongOh Kim**

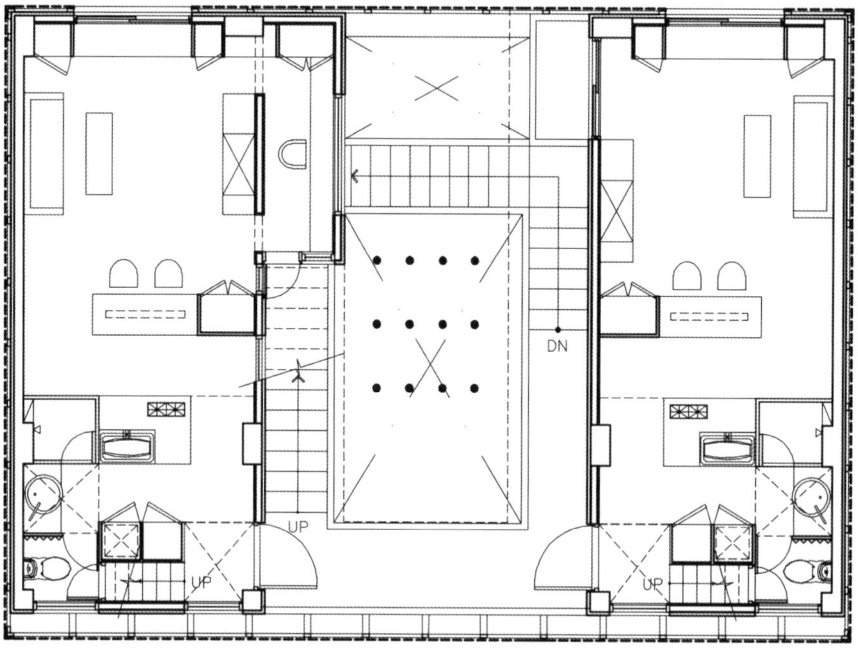

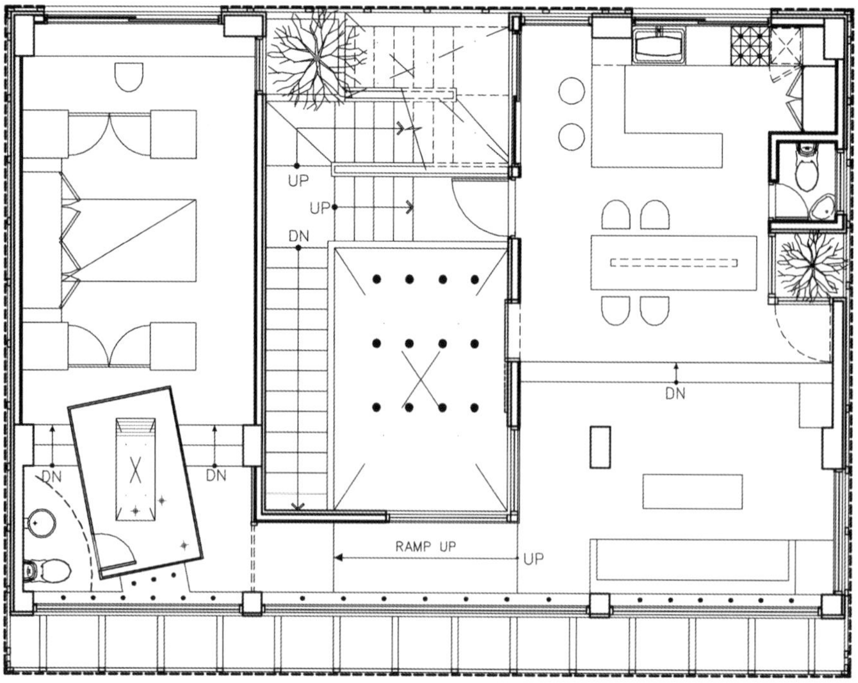

该项目是位于首尔市附近的北村，指向文化有些像汉城城墙。至今许多韩国风格的房子仍然存在，与许多普通家庭的住宅房屋一起发展。因此，它是村庄中的建筑，在现代化进程中与高层建筑相违背的建筑。这是一个现代化进程中不可避免的困扰，是一种新旧文化交汇缔造的房子。

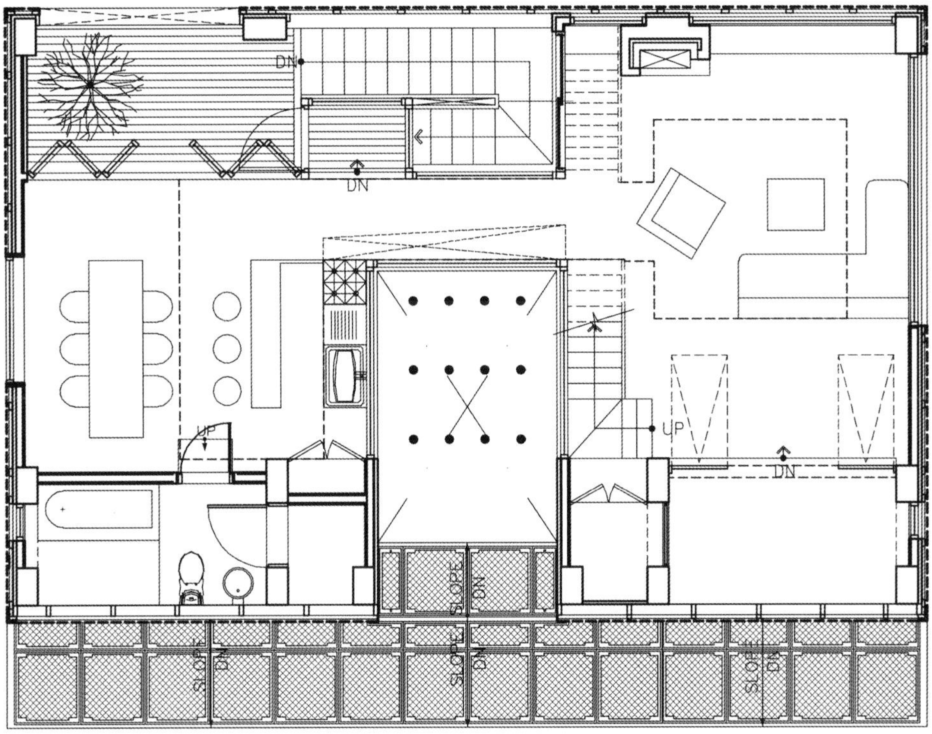

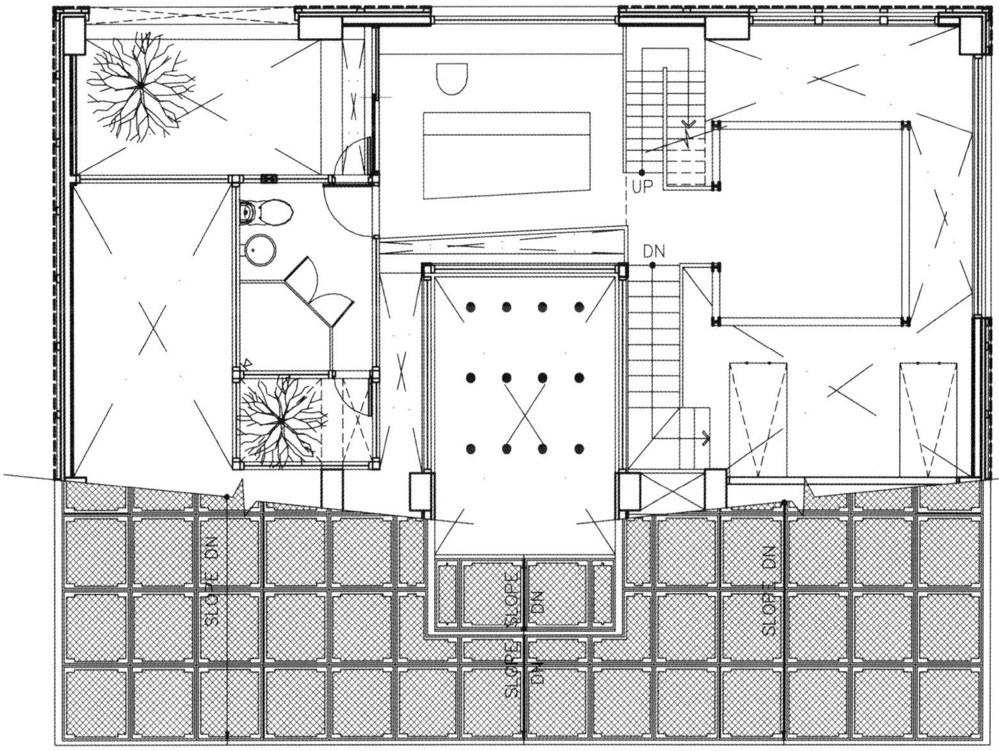

历史和现代的住宅空间通过组合元素继承对现有花园的认识。作为一个空间元素，满足了人们功能上的需求。在这房子里，每个家庭都拥有独立的花园。作为通道和外部空间将是一个城市管道，成为连接城市与建筑之间的通道。

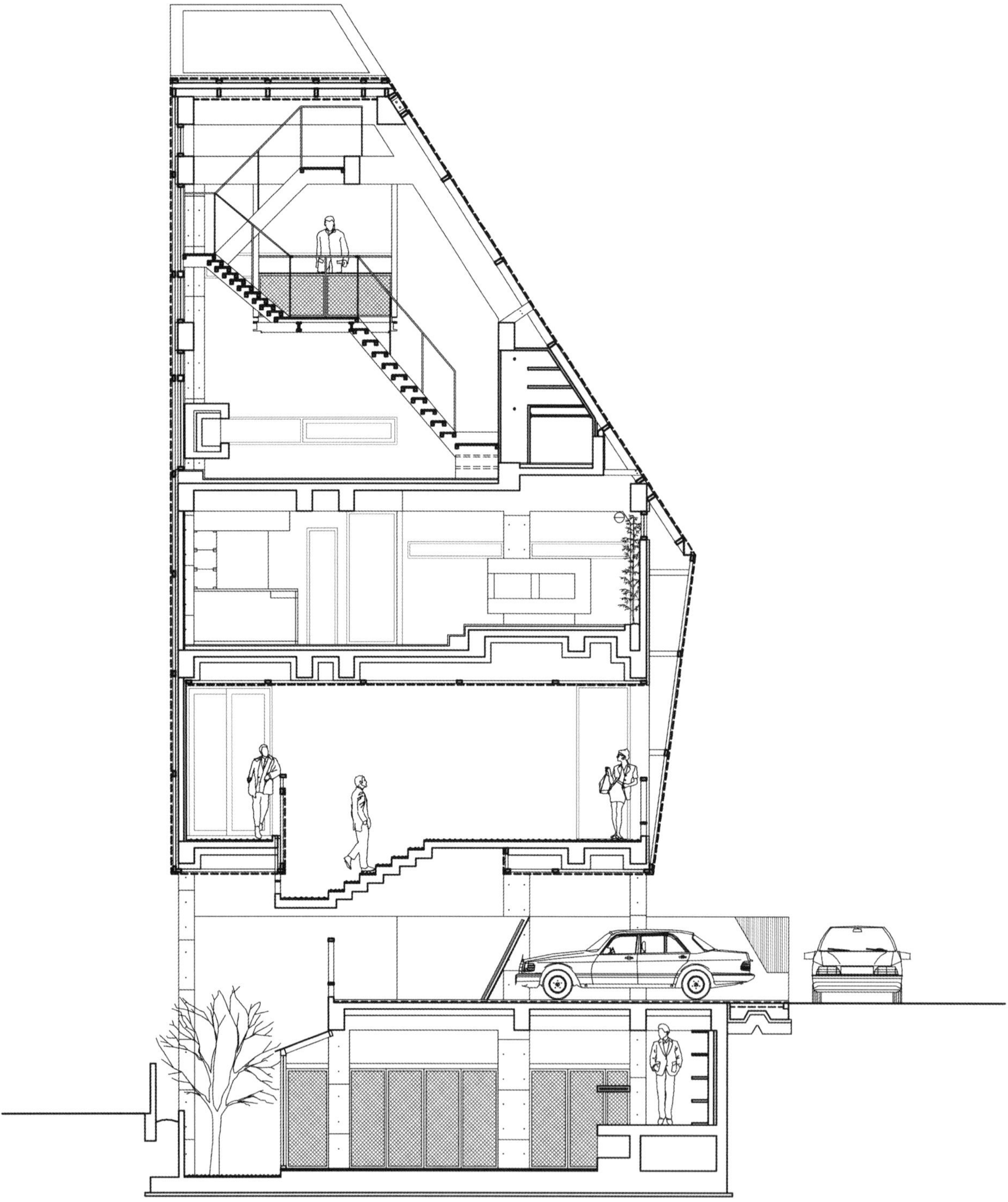

通过传入与导出的半透明金属板，光、风，每个相邻的房子包围院子，形成声效的景观，从而形成内部安静的空间环境。目的是：过滤周围的景观，与半透明的外立面相邻。透过模糊的景观，韩国风格的房子外观是可见的，与南山的汉城塔和夜景相融合。城市与建筑被赋予了强烈的呼应感，最后成为这种半透明的大规模的建筑。目的是：群体建筑之间巨大的质量比较，以协调与稳定各个方面，其余由韩国式住宅形成景观，形成新的城市文脉，来对应这一变化。它失去了阴影部分，有权享受阳光，保护文化遗产。由室内种植到外围，它可以被看作一个生态平衡，以及整个半透明/不透明的群体建筑为对象，并形成统一韩国风格房子的外立面装饰。

我瞄准了大自然的美好，目的在于积极传播这种居住思路，即丰富的居住空间与大、小房子错落在一起。通过建立公开的私密的生活方案，就像淋浴、浴缸、洗脸台等隐蔽的设施，他可以摆脱私密性，并可能呈现生活中的每一点。随着感受到的深化，在一种本能的环境中，所有行动成为一个视觉的关注对象。因此这是一个有趣的设想，其中存在的历史和故事总是有人知道的。这些方案虽小，但可以说这是大的。虽然是小空间，但形成不同的大的空间感觉，减少建筑开支。建造费用的节省，是成本管理的有效体现。尽量减少使用像混凝土材料的楼板，墙体去除装饰材料，预制夹心板和流程简化，这些都极大地降低了成本。V

Project: Slit Court
Design team: EASTERN design office + HOJO STRUCTURE RESEARCH INSTITUTE
Location: Kyoto Japan
Building area: 440.12 sqm
Photographer: Koichi Torimura

The essence of this architecture lies in an inner court.

The site is at Sumizome, Fushimi, Kyoto. It is a small town with a long and distinguished history, situated south of Tofukuji temple. Facing to south, the building is built on a gentle slope.
It is a collective housing of five stories. Two tenants occupy the front and back side of the first floor. There are four units on the second to fourth floor and two units on the fifth floor.

The site of a Japanese house is generally small and a narrow lot faces a narrow street. Even in a historic town, it happens that such a narrow street is turned to a main street to be the transportation route for cars, then it is no longer a place where town people once shared the joy of living together.
Therefore in our architecture, we make the inner living space a place to protect what matters in life. We try to protect sky, land and light inside the building.
In this occasion, we are asked how the way should be that will lead us to an important place.

One's beautiful life shares the destiny with a town. And the first place where people encounter the outside world is a path.
We, therefore, try to adjust the form of the path to the history of the town, the clients' way of living, and the way of existence of the town people living there. The resident of the architecture built on the basis of such notions declares that this is the way his beautiful life should be.
We made architecture with an inner court. This inner court redefines how the form of "people, town and path" should be in Kyoto.

Making a place which is slightly closed to public and hidden from others a serene space. In this place, the spirit of people is playing fully. We created such a place in the center of the building.

Slit Court

项目名称：**Slit Court**

设计团队：**EASTERN design office + HOJO STRUCTURE RESEARCH INSTITUTE**

项目地址：日本京都

建筑面积：**440.12** 平方米

图片：**Koichi Torimura**

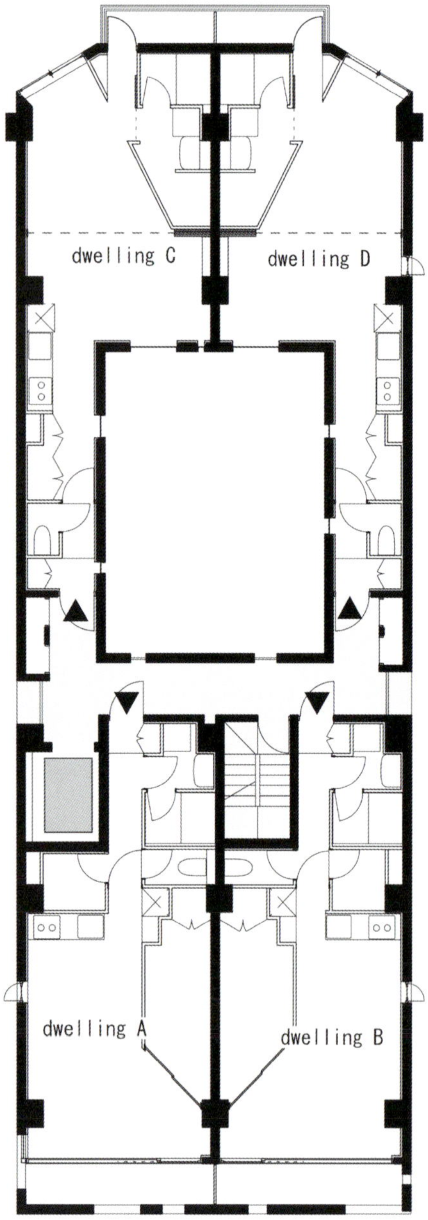

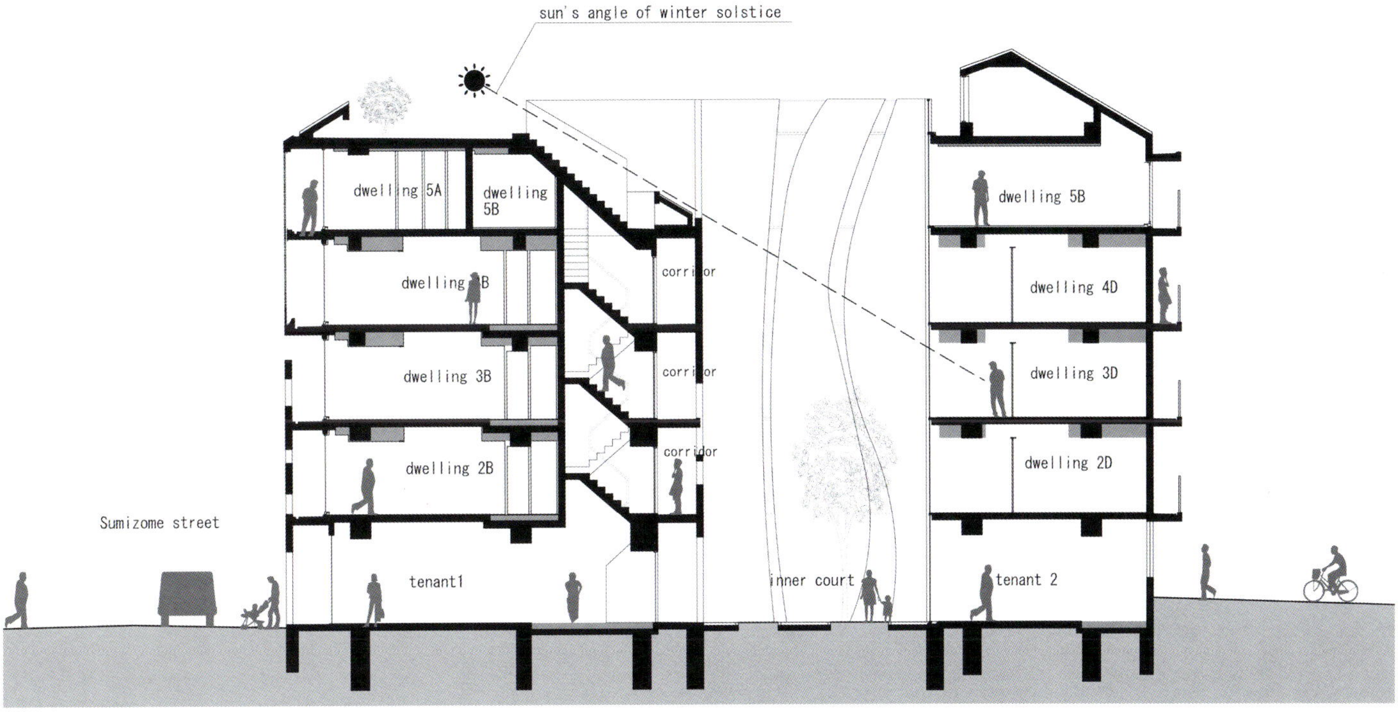

该项目位于日本京都。这个建筑的本质在于内在部分。这是一个有着悠久历史的小镇，建筑位于南部的Tofukuji寺。大楼建在南部的缓坡上。这是一个五层的住房，两名租户占据了一楼的正面和背面，分别在第二至第四层和第五层。

日式住宅基地一般很小，窄窄的街道面临很多问题。在一个历史悠久的城镇，这样的狭窄街道变成了主要街道，而且有机动车道，那么它已不再能引起原来在那里居住过的人们的快乐回忆。因此，在我们的架构下，使我们的生活空间内的地方得到保护。我们试图保护建筑物内的天空，土地和阳光。在此，我们思考的方式是将导致我们改变的原因。

一个人的美好生活可以改变城镇的命运。我们尝试调整道路的形式，居民的生活方式，以及城市人生活在那里的存在方式。关于这些概念的基础上建立起来的建筑被称为，美好生活的载体。

戈们做一个内部庭院的大楼。这种内在的力量重新定义如何将“人，城市和道路”合理化。这是个稍微封闭的区域，公众和其他隐蔽的地方构成一个宁静的空
]。在这个地方，人的精神得到舒展，我们创造了这样的建筑。V

Project: XTEN Architecture
Architect: XTEN Architecture
Location: Santa Monica, California, USA
Building Area: 820 sqm
Photographer: Art Gray Photography

The Diamondhouse is a sound studio and office extension to a house located deep in a canyon, against a severely sloping hillside, with minimal access and little space upon which to build. Direct sunlight reaches the site for only a few hours a day. The geotechnical condition is challenging, requiring 30-foot caissons to underpin new walls and foundations. A complex web of regulations governed the height, width, depth and specific relationship to the retaining walls needed to build the project. Given these constraints, a multifaceted architectural strategy was developed for the small building. First, a base building geometry was developed to conform to the hillside and required codes while maximizing the interior spaces by extending them into adjacent sideyards. Like a rock placed in a small pond, the addition is carefully placed between the existing structure and an imposing hillside to inflect the landscape and create exterior programmatic spaces around it where none could exist before. The building geometry also conforms to the interior program as a corner of the upper floor flares out to accommodate a writing desk built into a north-facing window and a series of wall planes fold up and over the building to create a rooftop railing and enclosure. Next, a building material system was developed to both relate the new extension to the natural landscape and to reduce the visual scale of the building. The faeade pattern was created from natural elements taken from the canyon site; abstracted, scaled and arrayed across the building in 3d modeling programs. Many patternsand scales were investigated in this manner, from which several full size prototypes were made by laser cutting and cnc routing aluminum, zinc, steel, plywood and cement panel samples.

The 3/8" thick Swisspearl fiber cement panels shown on the final construction were selected for both the way they relate to the masonry site walls and original structure, and for the fragility they impart to the otherwise hard edged and programmatically driven geometry of the building. Daylight brings out the etched lines, irregular edges and delicate qualities of the cement panels, which in direct sunlight appear as thin as a ceramic vase. While in the evenings the panels become flattened and the overall volume emerges, appearing as a patterned and perforated lantern illuminating the canyon.

XTEN

Architecture

项目名称: **XTEN Architecture**
建筑设计: **XTEN**建筑
项目地址: 美国加利福尼亚圣莫尼卡
建筑面积: **820** 平方米
图片: **Art Gray Photography**

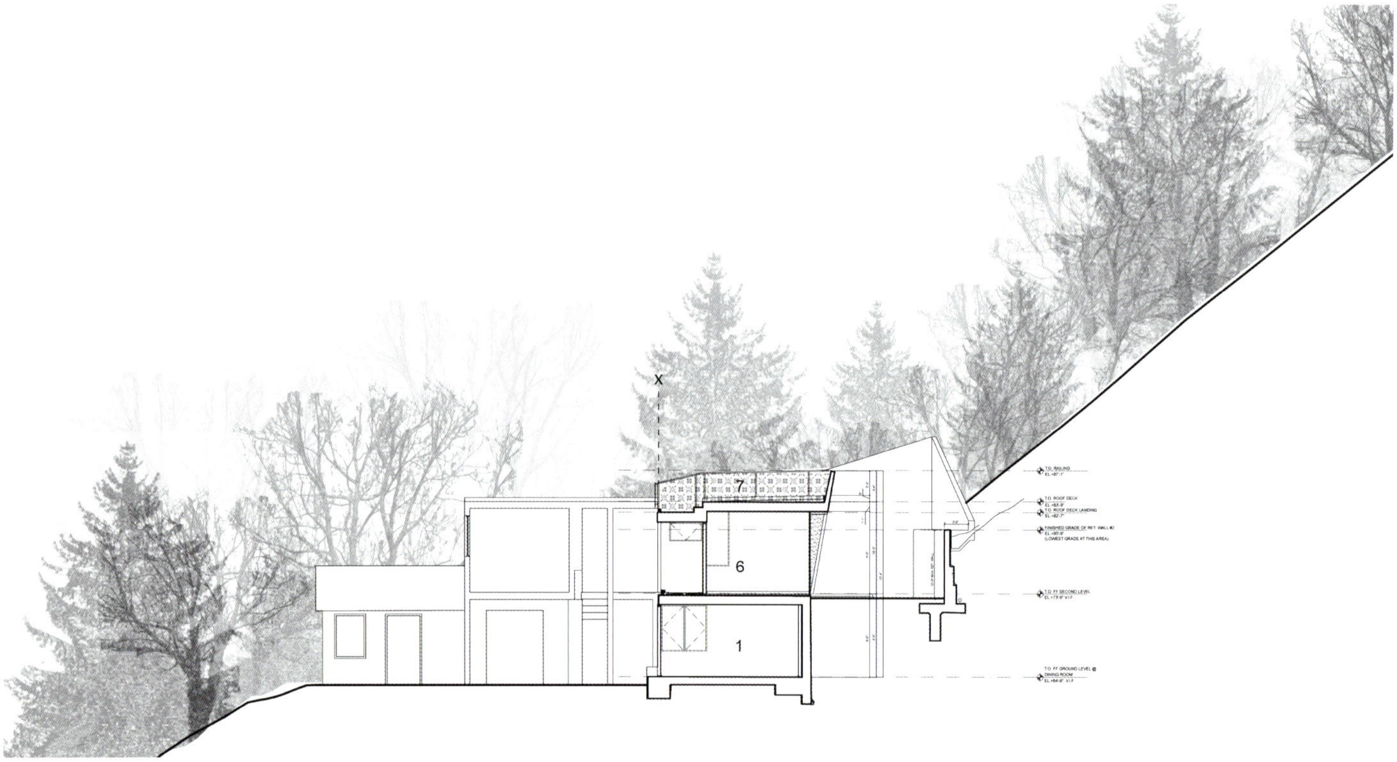

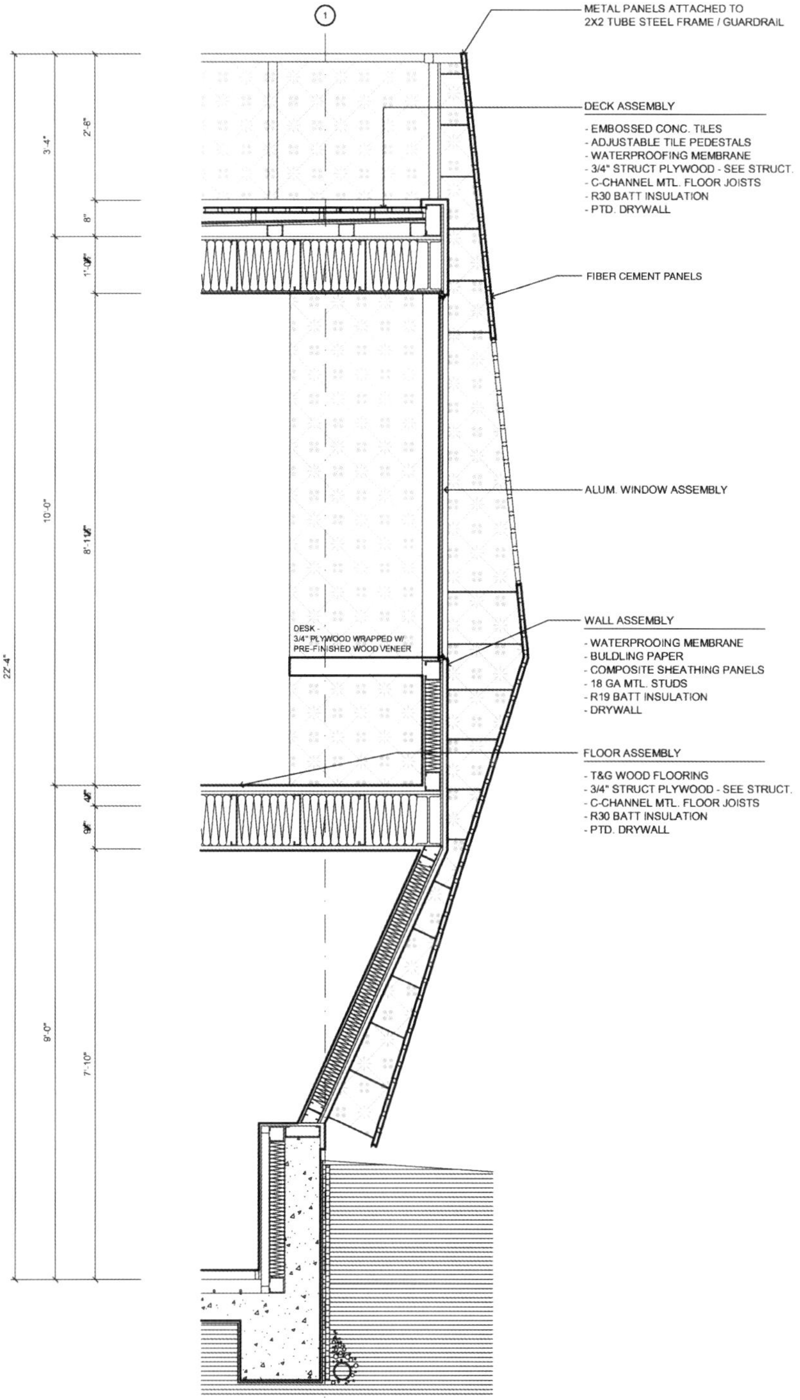

该项目是录音室和办公室的延伸，地点位于峡谷深处的一个陡峭的山坡上。小屋内，阳光直射只有短短数个小时。岩土条件是具有挑战性的，需要30英尺的地基，用以支撑新的城墙建筑。错综复杂的地面环境，高度、宽度、深度和相互的关系，迫使施工时需要建立新的墙壁。

首先，基地建设几何形是为了顺应山坡，最大限度地利用地形，而延伸到相邻的内部空间，就像一个小池塘里放了一个岩石，除了小心地放置在现有结构和气势宏伟的山坡上，还需创造空间令周围很好地与之融合的景观。该建筑的几何形状符合内部结构，作为上层楼角可容纳到北面向窗口的新建筑。

其次，建材系统的开发既涉及新的扩展的自然景观，还要减少建筑物的视觉范围。建筑立面格局是抽象的，是利用整个建筑规模和排列在三维建模软件上的。通过这种方式，对许多尺度进行了调查，由这几个尺寸原型和数控激光切割路，这些都由铝、锌、钢材、胶合板和水泥面板样品制成。

最后，使用显示板的方式都涉及到现场砌筑的墙壁和原有的结构纤维水泥而建造。日光带出了蚀刻线，不规则的边缘和细腻品质的水泥板，水泥板在阳光直射出现像陶瓷花瓶一样的透光性。在晚上，面板变得扁平，作为一个图案和穿孔彩灯照亮了整个峡谷。V

Project: Mirante Do Horto House
Location: Mirante do Horto Condominium - Horto Florestal - Paulo – SP - Brazil
Architect: Flavio Castro
Building area: 300 sqm
Photographer: Nelson Kon

The design of the residence came from a basic premise: maximum use of land for a huge program. Thus, the strategy was to locate specific areas and vertical circulation In a way that the rest of spaces enjoy extreme flexibility and a Wide visual range, profiting the structure and volume.

The space created inside, in fact, is what remains after adding three small "boxes". Dialectic interdependence between home theater, kitchen and stairway is established in the residence and demarcates a territory, as seen illustrated in the sketch next to this text.

The service, adjacent to the garage, is four feet below street level. Above is the container with the other functions of the house. A shadow separates them.
The shadow on the outer part translates into light in the inner space. The range allows the lighting and ventilation of the environments below and gives an extreme light visual to the massive upper volume.

The structure is the protagonist in the conception. Accesses through this gap were distributed as independent elements of the geometry of the house.

A container rests on two "bars" and stands out in access and uses of their immediate surroundings. The prismatic volume contains the program of the residence on two floors plus a roof garden. The container openings are extremely flexible in the front and back facades, while quite tenuous on the side walls. This is a consequence of the immediate surroundings because the common practice of the neighbors is constructing on the border of their private grounds. Because of this, the program is located in the perimeter side providing a more flexible interior space and visual communication in height.

This transition provided by two metal beams allows the floor of the garage to spread to the maximum boundary of the site. Maximum occupancy without affecting ventilation and lighting guaranteed exactly by that range.

Connecting the house from top to bottom, the right wall next to the staircase inside the house offers vertical connections to the exchange of fluids between the four height levels, while providing space for a different event in each. On the garage floor, superior and top terrace, it generates the studio, office and support respectively, while in the main floor a large aquarium turns that wall in a prism.

The house was designed by a succession of free horizontal slabs with strategic holes. In these 8x11 meters plans, these holes allow vertical circulation and the entry of zenith light. Natural light enters next to the water tank, passes through the upper floor and reach the social floor.

Besides being a viewpoint to the city, the roof garden contributes to the sustainability of the residence because it helps cooling the area immediately below, the dorms.

mirante

o Horto House

项目名称: **Mirante Do Horto HousE**
项目地址: 巴西圣保罗
建筑设计: **Flavio Castro**
建筑面积: **300** 平方米
图片: **Nelson Kon**

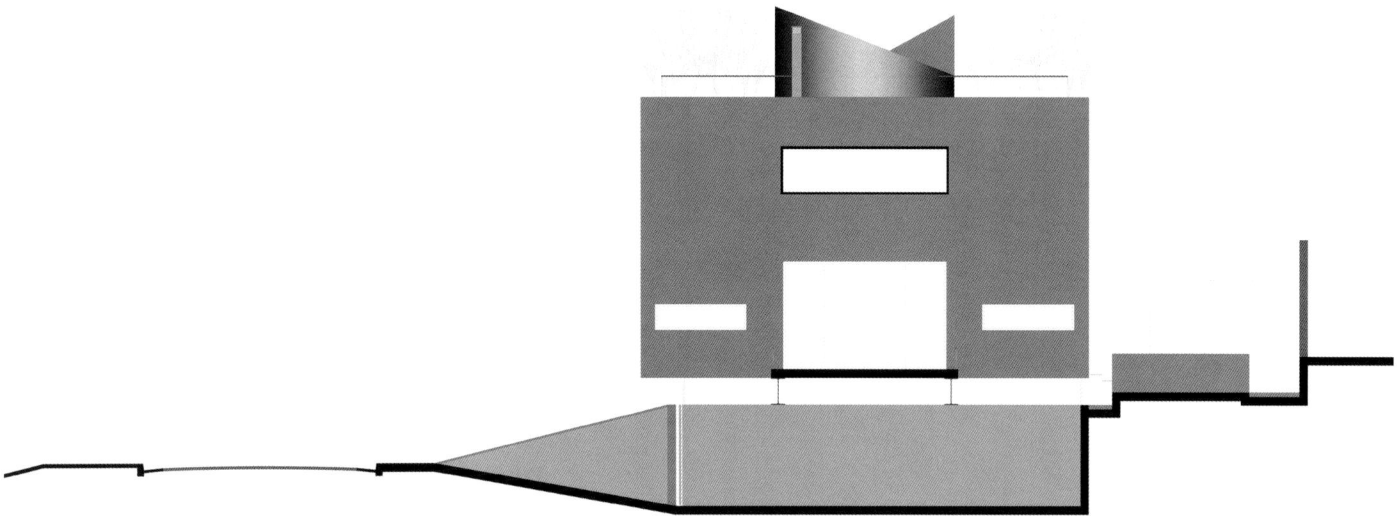

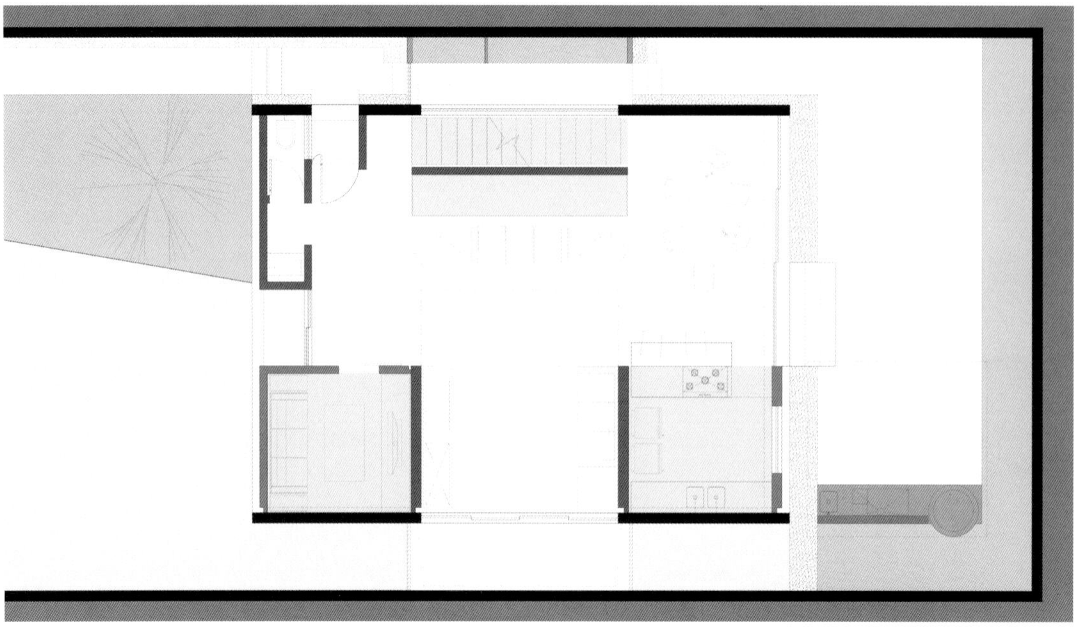

该住宅设计的基本初衷是：最大限度地利用土地。因此，该项目的定位是：其余的空间享有极高的灵活性和极大的可视范围，结构和数量等具体方面和垂直布局。在内部创建的空间，其实是后加入的三个小"盒子"。家庭基础设施，厨房及楼梯相互依存是建立在一个地区居住的基础。外部分的空间环境、允许范围下的环境照明和通风，都给出了一个极端的光影视觉上的差异明显。这种差异是独立的房子在几何元素上的差异，其结构是概念上的主角。

这所房子是由一个很成功的设计团队设计的。在这些规划中，允许光线由天顶垂直洒下。自然光线进入旁边的水箱，通过上层的通道到达地面。因为，它有助于改善该地区的寒冷。屋顶花园有助观赏休闲，使室内的温度得到良好的保持。

Project: White Curve
Building area: 600 sqm

Reminding us of scenes shape the future of architecture, is so dramatic, unexpected curves lack indoor and outdoor use, the whole room full of elegant mystery.

Great window design, a broad perspective, to the surrounding environment glance, all overlooking the height determines the height of thinking, refreshing feeling from the heart. The open interior space for water-like rhythm, such as around the smart.

White Curve

项目名称: **White Curve**
建筑面积: **600** 平方米

恍若未来建筑的造型，是那么的引人注目，室内室外不乏出人意料的曲线运用，让整个空间充满了优雅的神秘色彩。巨大的开窗设计，广阔的视野，让周边的环境一览无余，俯视一切的高度决定了思维的高度，心旷神怡的感受发自内心。室内空间的开敞让韵律感如流水般环绕灵动。

In this brief overview, there is a hierarchy of topics and we have not exhausted in an argument each. Just chose themes that struck me as timely and wrote about them. Thus, only the text accompanying the work, because I still believe that the chart is the primary tool for understanding the entire architecture presented.

Four virtual axes perpendicular to organize public street spaces of different nature within a wrapper. In this scheme we see an empty hub that links the two halves and a third element to the bottom ends of the patio enclosure later.

While reading the first volume in a front facade, the building soon awakens certain duplicity in being composed by planes or volumes. Le Corbusier deposited in shadow and light as a result, much of your concerns about architectural compositions and I think she is, in this case, the driver of much of this discussion.

Finally, the contour of each face exposed to the light and the distances between the vertices and edges of the parts in question makes us realize that this is a relationship between plans. I'm saying this not as a criticism of the volumetric composition, merely clarifies the process that was adopted in this house and some other positions that have left there. Since the author assumes the actual result of his work, such as Aalto and Siza take in relation to light, I believe that architecture can be built very well reasoned.

At the center, a long hollow shaft movement makes the visual connection throughout the longitudinal extent water tank that brings natural light and mark the central axis as an exception. While the transparency of glass lets in sunlight, natural depletion afforded by two shutters together with this cover eliminates the hot air that rises through its lower density.

The openings are projected to be about to heat stroke and the degree of privacy desired by customers. Side neighbors, the gable is almost blind, because we only have vertical circulations and swimming, while others are quite transparent in the two sides. The logic of regular openings greatly facilitated the structural calculations and execution, even as a constructive procedure could be adopted for almost all of the building.

HOUSI

g Pernambuco

项目名称: **Housing Pernambuco**
项目地址: **Guaruja / 巴西**
建筑设计、园林绿化、照明: **Flavio Castro**
建筑面积: **500 平方米**

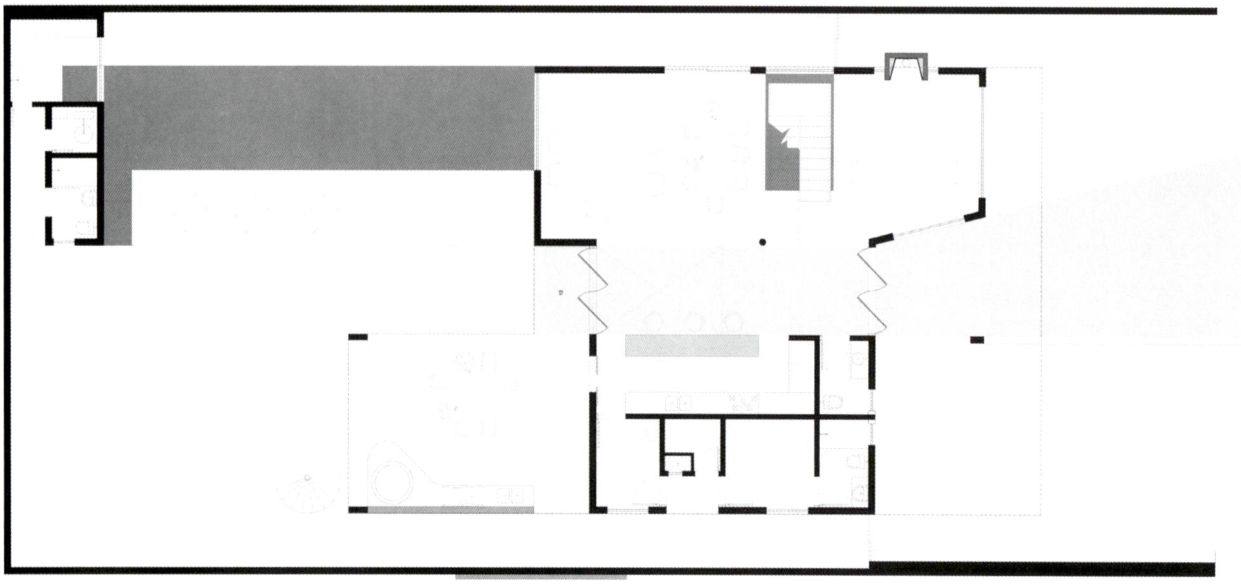

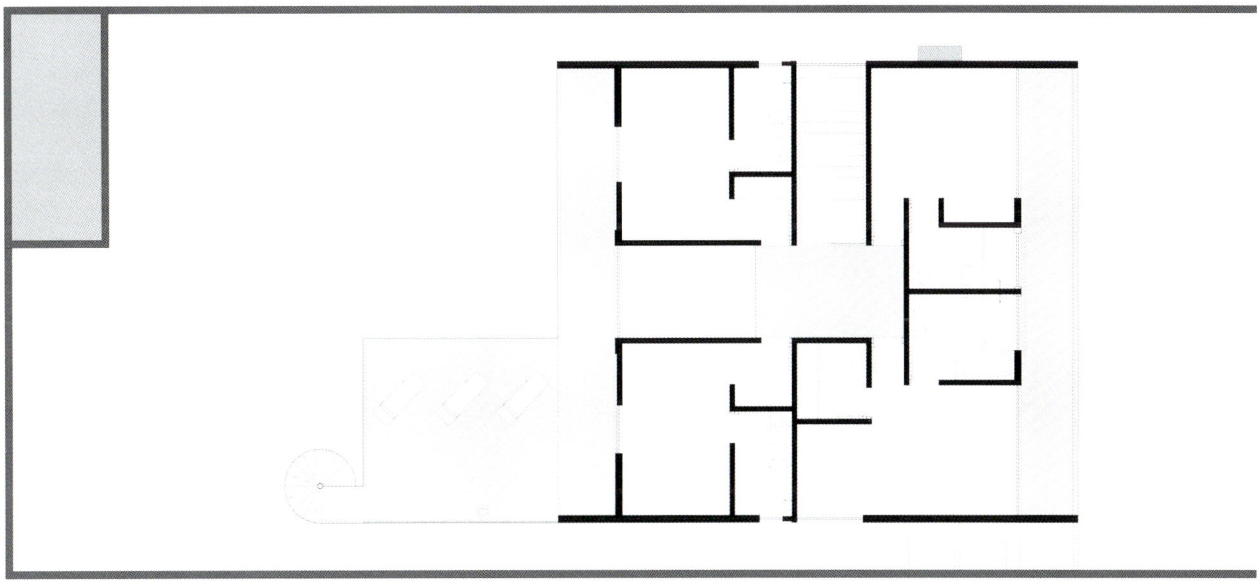

在这个简短的、有层次的项目中，我们没有把每一个细节参数要求得苛刻，只是在选择的主题上，可算是费尽心思。因此，文本所附的工作只是一部分，图是了解整个建筑呈现的主要工具。四个虚拟轴垂直的包装内组织不同性质的公共街道空间。在这个方案中，我们看到一个空的枢纽，连结两半，天井底部的外立面结束于此。

在领略开篇之作后，建筑的内部被立即唤醒，建筑在沉积的阴影和光线的作用下，形成了视觉的重点。此情此景，给人很多的启迪。

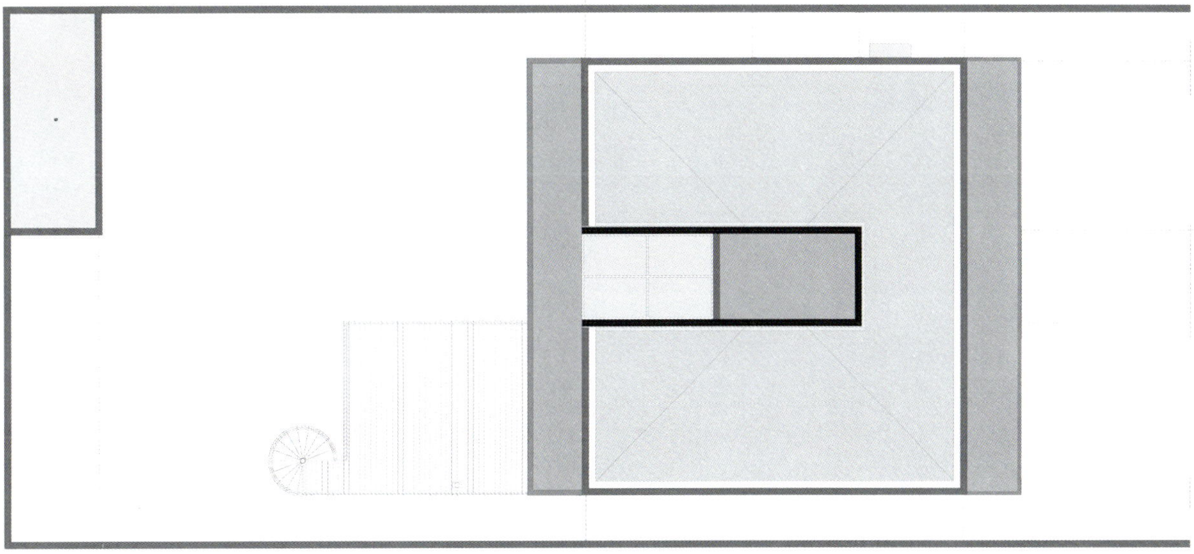

每个人的脸庞暴露在光线之中，顶点和边沿的问题使我们认识到这是设计之间的关系问题。我的意思不是作为一个组成的部分批评这一点，只是明确了这所房子的整体性，由于作者认为，他的工作如阿尔托和西扎一样，根据实际相关情况采取行动。相信建筑可以建得很好！

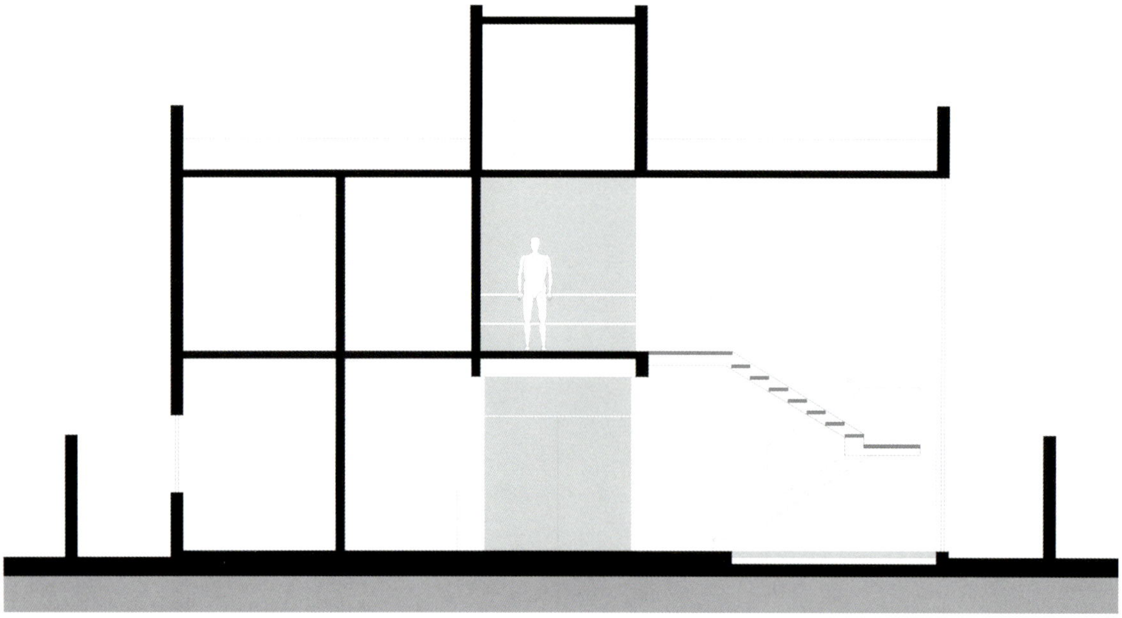

在中心，长的空心轴引导着视觉，连接了空间。自然光标记为一个异常的中轴线，虽然透明玻璃和百叶窗共同消除了热空气，但其密度还是不能阻止热空气上升的。

不开窗户的原因是：户外的温度估计会让人中暑，另外也不能忽略保护房子主人的隐私。同时，侧面还有邻居，山墙几乎是毫无遮拦的。我们只有观看垂直环流和游泳的时候，才打开窗户让双方完全没有阻隔。V

Project: Acapulco House
Location: Brazil
Architect: Flavio Castro
Building area: 334.29 sqm
Photographer: Flavio Castro

The house is composed of 6 squares grouped in pairs. Within this compositional rule, two subtractions were made at opposite ends together to form two venues: the garage and playground. The various relationships that they establish with their empty interiors were offered by the unevenness of the floor, transparency and accessibility. In the sketch below (sketch of No. 6 - very important), we can see the outline where references to which I clearly perceive an empty hub that links the two halves and a flag to the bottom end to define the backyard.

Within this dynamic of squares, four virtual axes perpendicular to organize public street spaces of different nature within a wrapper. At right, the service is organized linearly and goes unnoticed to the visitor while composing volumetric important role. You are in a strategic location with respect to insolation and functionality.

Left, the succession of rooms and leisure facilities provides the tenant with a space width 5.50 m tall and huge transparency in parallel to the street facades. Dita transparency reaffirms the continuity of space bounded only by the huge white planes with residents suggest that a full ownership of your home.

A single discontinuity happens to the access stairs to the bedroom floor. Metallic self-supporting, acetic and colorless, it cuts the sequence of rooms at one point creating a visual filter between the public street and private backyard.

At the bottom left portion of this flag that ends the succession of such environments. Separated from home by the range of pool and deck, creates a tension between his own house and volumetry. This space interval quality wins in the duality between "seeing and being seen, completely dissolving the notion of attachment.

Inside me a little now in the dynamics of the bond that exists perpendicular to the axes mentioned earlier. The TV room and recreation are liable only for union with the physical opening of the doors that separate them initially. You could say that these two environments mixed up in favor of a more generous social interaction and changeable. Almost as if the higher volume that houses the dorms had migrated to a higher elevation to promote coexistence and for the establishment of a cross-ventilation as shown in the sketch below.

The openings are projected to be concerned about the touches that occur between the geometries that form the house. Are stress points that deserve special attention. Neighbors side, the gables are not punctured, while others are quite transparent in the two sides. The series of openings in the facade right side indicates a more bureaucratic program that is united by a pronounced roof slab reinforcing the linearity of the upper block. Even the volume of the water tank following the academic adopted from the outset that segregates the whole into parts, as well as a few centimeters away from the lower deck in plan, a distinct chromatic value was assigned to this "box" in spare time.

aulco House

项目名称: **Acapulco House**
项目地址: 巴西
建筑设计: 弗拉维奥卡斯特罗
建筑面积: **334.29** 平方米
图片: 弗拉维奥卡斯特罗

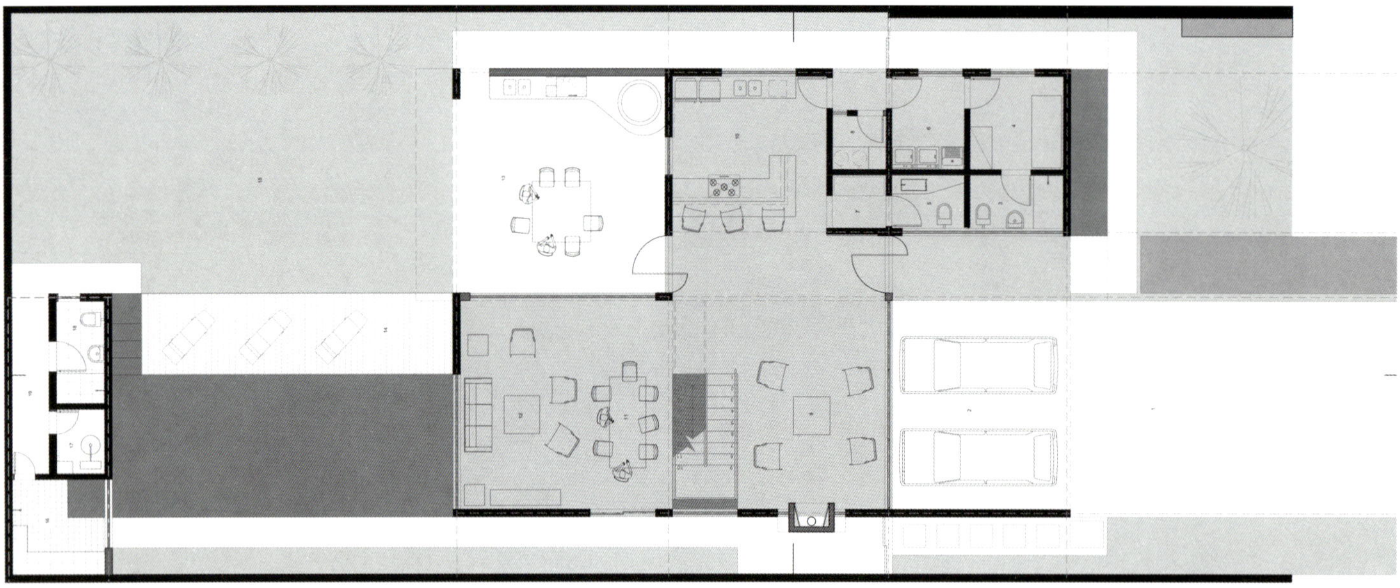

这房子由6格分组。在这个构图规则中，两个相反的空间在两端连在一起，形成两个场地：车库和游乐场。他们建立与他们的内部有着各种关系。在下面的草图中，我们可以看到其中提到的大体意思，清楚地感觉到一个空中枢纽，连结两半和一个标志，底端还有个后院。

EDY-0999

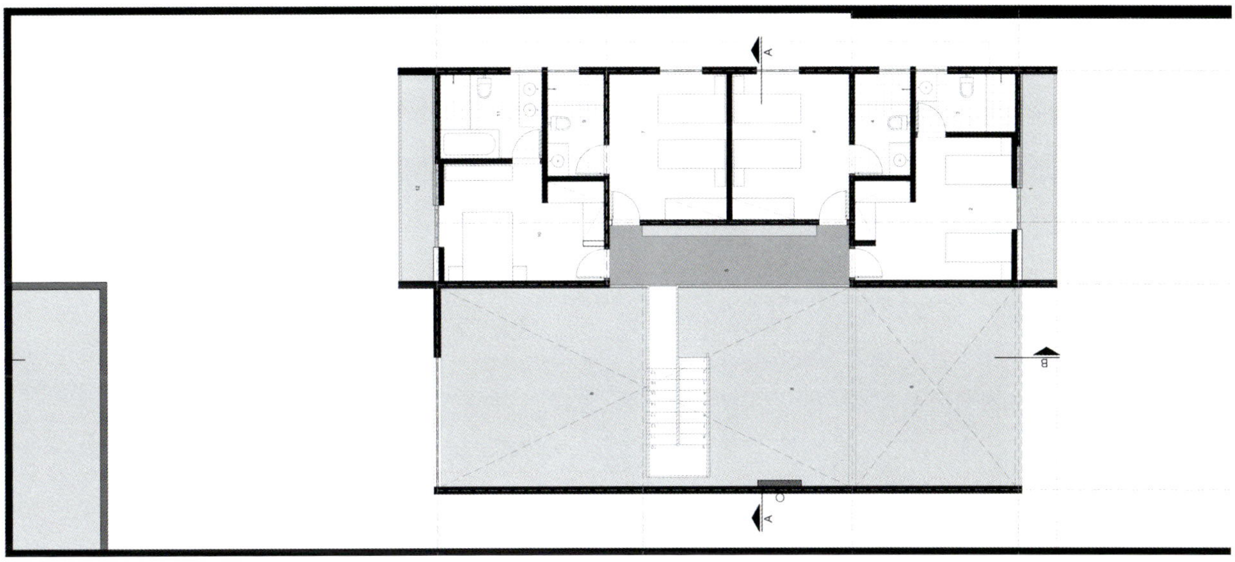

在这个充满活力的场地中，四个虚拟轴垂直空间内组织有不同性质的公共街道空间。在右边，是服务的空间，设置了房间和休闲设施，那里涉及重要的功能——晒太阳。这里提供了一个5.5m的平行于街道的巨大透明立面。建筑的透明度重申了空间的连续性，连续发生在一个单一的楼梯上，直到卧室地板上。它减少了在一个点序列房间之间的建立，公共街道和私人后院视觉过滤了私人后院。

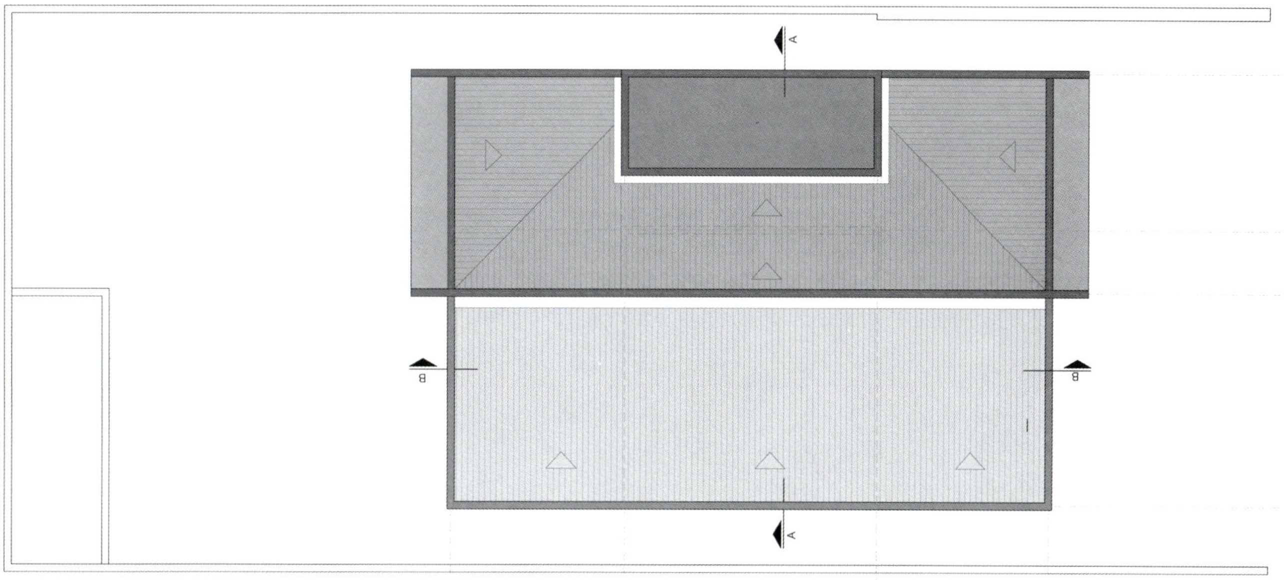

在这个标志的底部，结束了环境中沿袭过去的部分，从露台到游泳池这些范围中，创建一个能让他自己从紧张中得到松弛的空间，在这个空间之间的间隔界于发现与被发现的抽象概念。我心中存在了前面提到的垂直轴的动态空间，电视和娱乐仅仅是一扇独立的门，他们是开放式的空间，你可以说这是混搭环境的互动，现在倍受青睐。仿佛高层公寓已经迁移到了高海拔地区，促进沟通和那种草图所示的相互联系之感。

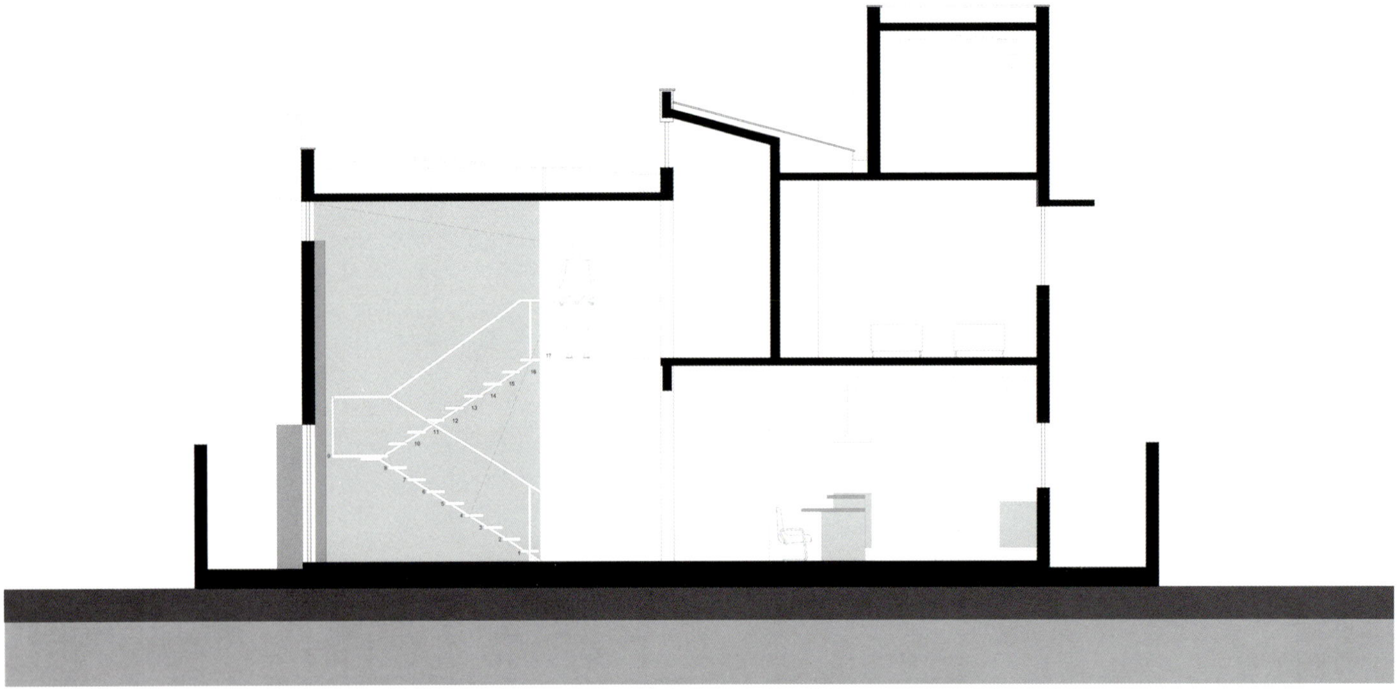

房子的主要问题：开间的设计将涉及有关的几何构成问题。这个关键点值得特别注意。邻居方，山墙没有打通，而有些则是双方可见的。在右侧的门开方向表明其严谨的设计，它是由一个明显加强了对屋顶支撑的线性板构成。整个空间的组成部分是用水箱容积分离开了，一个独特的色彩在空闲处被填进了这个“盒子”的空间。

Project: House S
Architects: David Driesen, Tom Verschueren, Sacha Bratkowsky, Christine Loos
Location: Mechelen, Belgium
Building area: 238 sqm
Photographer: Frederik Vercruysse

House is a container, but with the increase of family members, children grow up, the original house seems more and more crowded. Mechelen in Belgium, one of the family is also facing this problem, how to grow the small children in a more comfortable environment at the same time they strengthen the "one roof" concept has become the focus of the parents. Elsewhere than move to a strange, parents tend to carry out alterations to the original building, in a familiar environment to create a new home. In the end, they go to Belgium dmvA penned the local design team, in full consideration 了 structure of the original construction needs with the owner, the designers of the bold De transformation, for the owners to create the one only Shuyu their "castles in the air".

Handling in appearance, designers both sides of its split into the design, they retained the original facade facing the street side, while the rear design for the building exaggerated shape. Front to back, from black to white, the conflict will be "home" highlights the need Wenxin Jing Yi. Frontages is still dark walls, brick roof composed of roughly a look around the houses and no two different designers so deliberately simple treatment is to create a hidden appearance, enhance the privacy of the family. The only bright spot in the entire street surface, is the design of doors and windows. Designers abandoned the traditional people in the concept of doors and windows are used with different sizes of holes in the aluminum folding doors and windows. Although the design of such a light shield, but a certain degree of sun protection. But also because such doors and windows designed to make building the internal space must rely on the design of the rear light support. In the night came, the house orange light through the rectangular hole on the aluminum windows and doors, transmission to the outside, like a carved lanterns.

Handling in appearance, designers both sides of its split into the design, they retained the original building external frontages stand on the back of the white space, transforming the focus of designers in the original three-tier structure were carried out on the Yibiao design. Changed the ordinary three-space design, designers will be converted into three different sizes respectively, the white cube, and the child's building blocks as normal, layer upon layer piled up. To the vertical development of the whole building, from the deepest depth of 17m to the bottom of the second floor of the 13m, and then to the top floor attic of 9 m, as a ladder-shaped, upward contraction. In order to maximize use of space, but also out of consideration of lighting inside the building, designers are not only designed in three large French window, is still in the two-story, three designed a glass-air walkway, so the sun can across from the huge windows, direct to the ground floor. The transparent air corridor is also the kids favorite entertainment; that separated but closely adjacent the design of treatment, creating a blind date for the house near each other, but separate spaces. Choice of colors in the furniture, designer white line resembles the choice to maximize the space look more spacious and clean.

Meanwhile, the designer has deliberately designed a loft balcony, for the whole family can enjoy a distant view of the place. To echo the minimalist white box, garden lawn from a combination of a square.

HOUSE S

项目名称: **House S**
建筑设计: **David Driesen, Tom Verschueren, Sacha Bratkowsky, Christine Loos**
项目地址: **Mechelen/**比利时
建筑面积: **238** 平方米
图片: **Frederik Vercruysse**

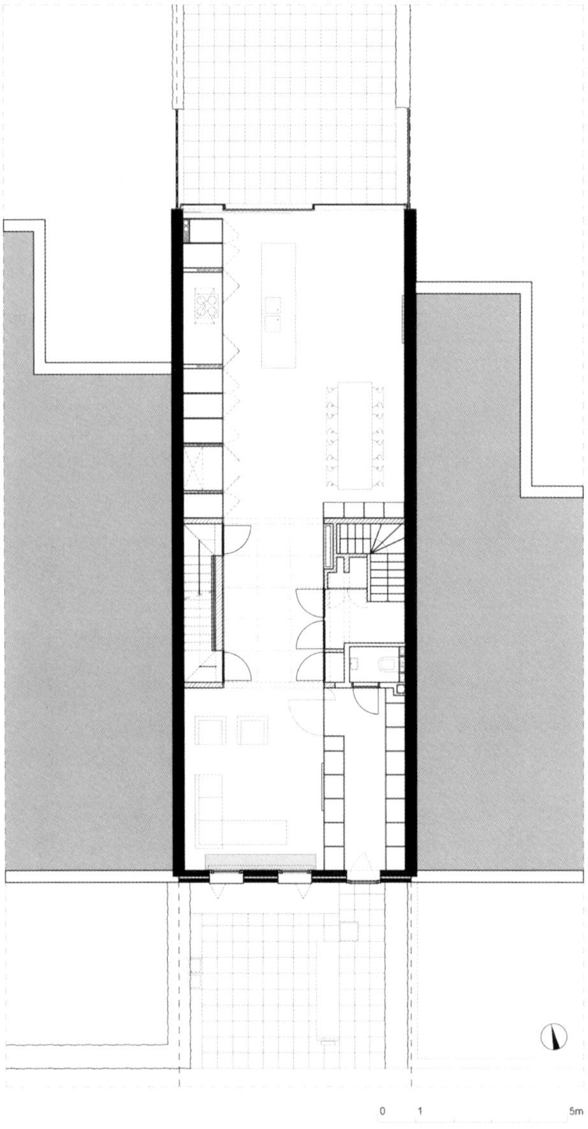

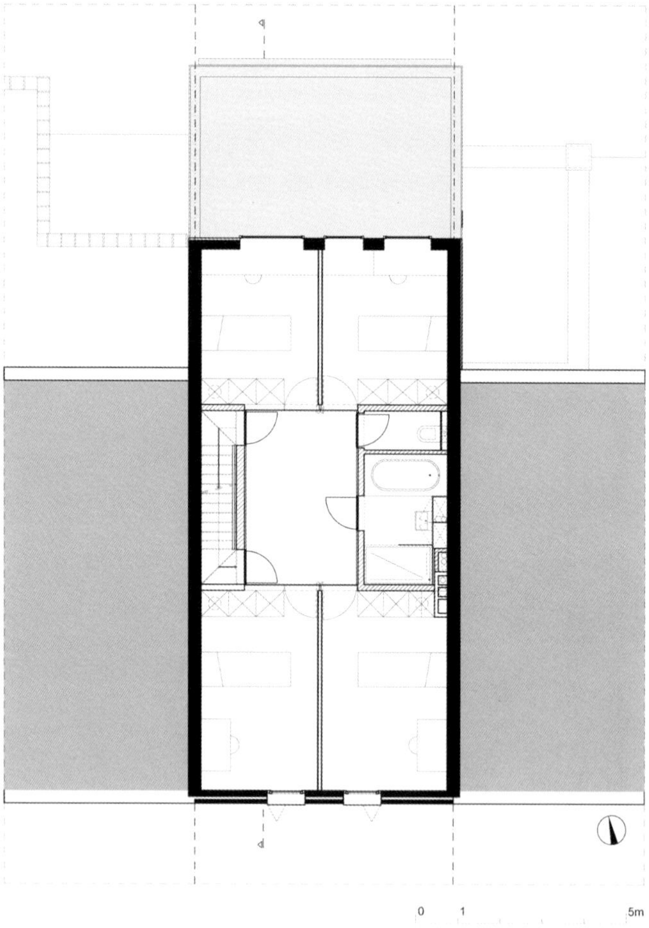

房子是家的载体，但随着家庭成员的增多、孩子的成长，原先的房子显得越发拥挤。位于比利时梅赫伦的某个家庭也面临这样的问题，如何为成长中的小儿女提供更舒适的环境，同时亦能强化他们"同一屋檐下"的观念，成为家长关注的重点。比起搬迁至陌生的别处，家长更倾向于对原建筑进行改建，在熟悉的生活环境中创造一个新的家。最终，他们请到比利时本土设计团队dmvA操刀设计，在充分考虑了原建筑的结构与业主的需求后，设计师进行了大胆地改造，为业主创造了一个只属于他们自己的"空中楼阁"。

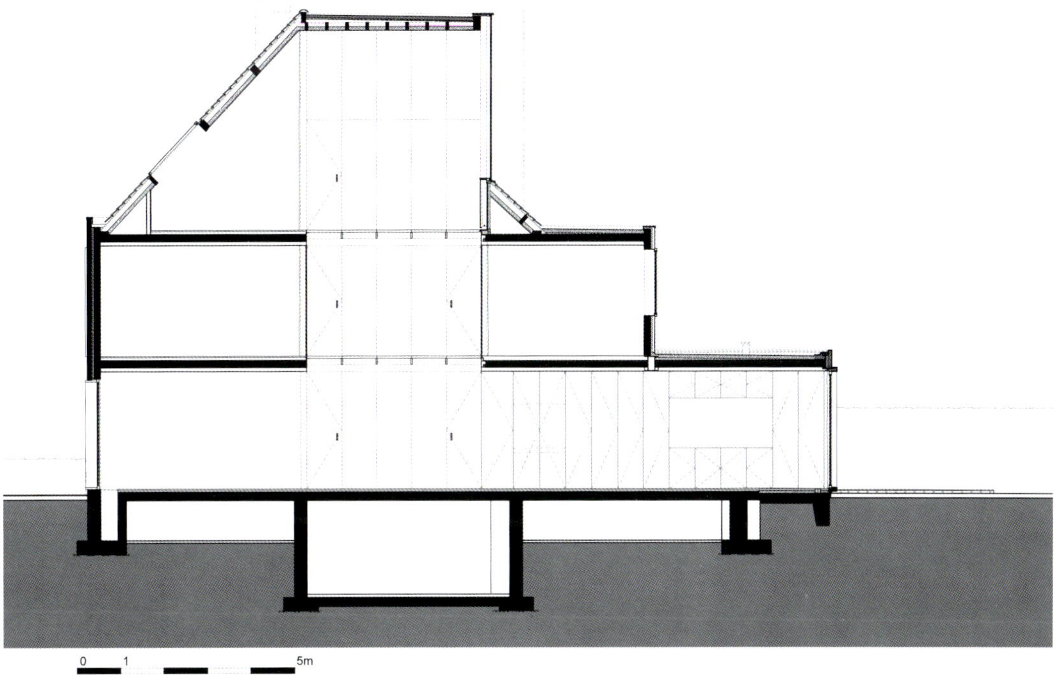

在外观的处理上，设计师将其分割成正反两面来设计，他们保留了临街面的原先建筑外立面，而为建筑的后部设计了夸张的外形。从前到后，从黑到白，在冲突中将“家”所需要的温馨静逸凸显。临街面仍旧是由深黑色的墙面、砖砌的屋顶组成，粗略一看与周围房屋并无二异，设计师这样刻意地简单处理，就是为了营造一个隐匿的外观，提高家庭的私密性。整个临街面唯一的亮点，是门和窗的设计。设计师摒弃了人们传统中门和窗的概念，均使用有大小不一的孔洞的铝制折叠门窗。这样的设计虽然遮蔽了采光，但是可以起到一定的防晒作用。同时，也因为这样的门窗设计，使得建筑内部必然需要依靠后部空间的设计进行采光支持。在夜晚来临的时候，屋内橘红色灯光透过长方形的铝制门窗上小孔，透射到屋外，像一个剔透的灯笼。

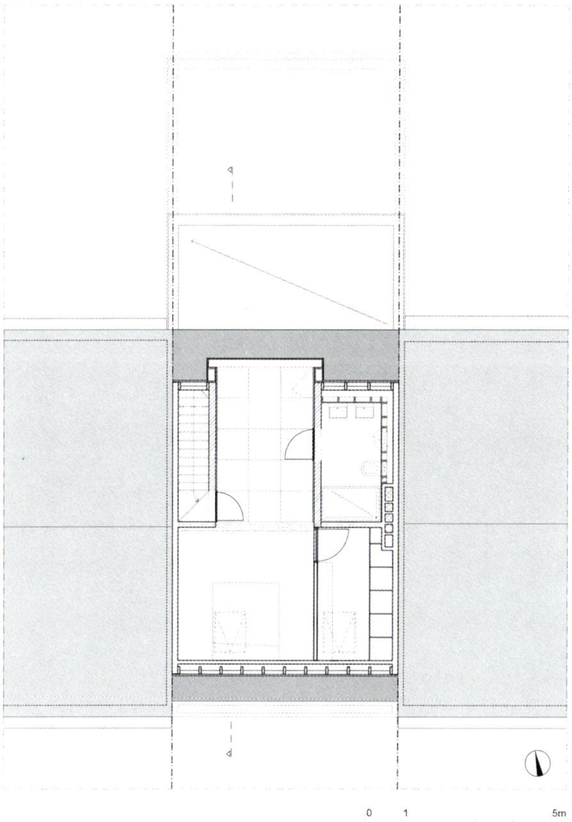

背面的白色空间，是改造的重点，设计师在原先的三层结构上进行了出人意料的设计。一改普通的三层空间设计，设计师将其分别改建成三个大小不一的白色立方体，并如同孩子的积木一般，层层堆砌起来。整个建筑向纵向发展，从底层纵深最深的17m到二层的13m，再到顶层小阁楼的9m，成一个阶梯状，向上收缩。为了尽可能地利用空间，同时也出于对建筑内部采光的考虑，设计师不仅在三层都设计了巨大的落地窗，还在二层、三层设计了一个空中的玻璃走道，使得阳光可以从巨大的玻璃窗穿过，直射到一楼地面。这个透明的空中走廊，亦是孩子们最喜欢的娱乐场所；这种相隔却又紧紧相邻的设计处理，为家创造一个相亲互近，而又各自独立的空间。在家具颜色的选择上，设计师亦多选择白色系，尽量让空间看上去更为宽敞与洁净。同时，设计师还刻意设计了一个阁楼阳台，让全家人有一个可以观赏远处景色的地方。为了呼应白盒子的极简风格，花园的草坪亦由一个个正方形组合而成。

Project: Villa Forest
Location: Brazil
Building area: 468 sqm

Project is located in a jungle in Brazil, building a great contrast with its architectural vision, to promote a harmonious, dynamic and static harmony. Building facades hanging beige stone facade has been with the surrounding natural color highlights the feeling of harmony, while the more prominent of the building mass and form a stronger visual impact.

Will be introduced to the water environment of the open-air swimming pool in the villa, the terrace has been extended to solid water, whether swimming in the water, or rest on the shore, people can be fully integrated into nature. Night, with lights next to the pool is opened, with the lighting of the reasonable intervention, but also lit up the sky brightened our mood.

Interior space, a lot of use, demonstrating the importance of local materials. Mad rough stone, mixed with the natural atmosphere. Prepared by the local reed curtains, the sun streaming into the room through the curtains dough, bits and pieces, patchwork. Extensive selection of furniture and the overall building form a strong contrast, creating a delicate slow life.

Villa Forest

项目名称：丛林别墅
项目地址：巴西
建筑面积：**468** 平方米

项目坐落于巴西的一处丛林中，建筑以其建筑视觉的极大反差，促成了一种和谐，一种动态与静态的和谐。建筑外立面米黄色的干挂石材，立面色彩突出一种与周遭自然和谐的感觉，同时，更加的突出了建筑的体量和形成更强烈的视觉冲击力。

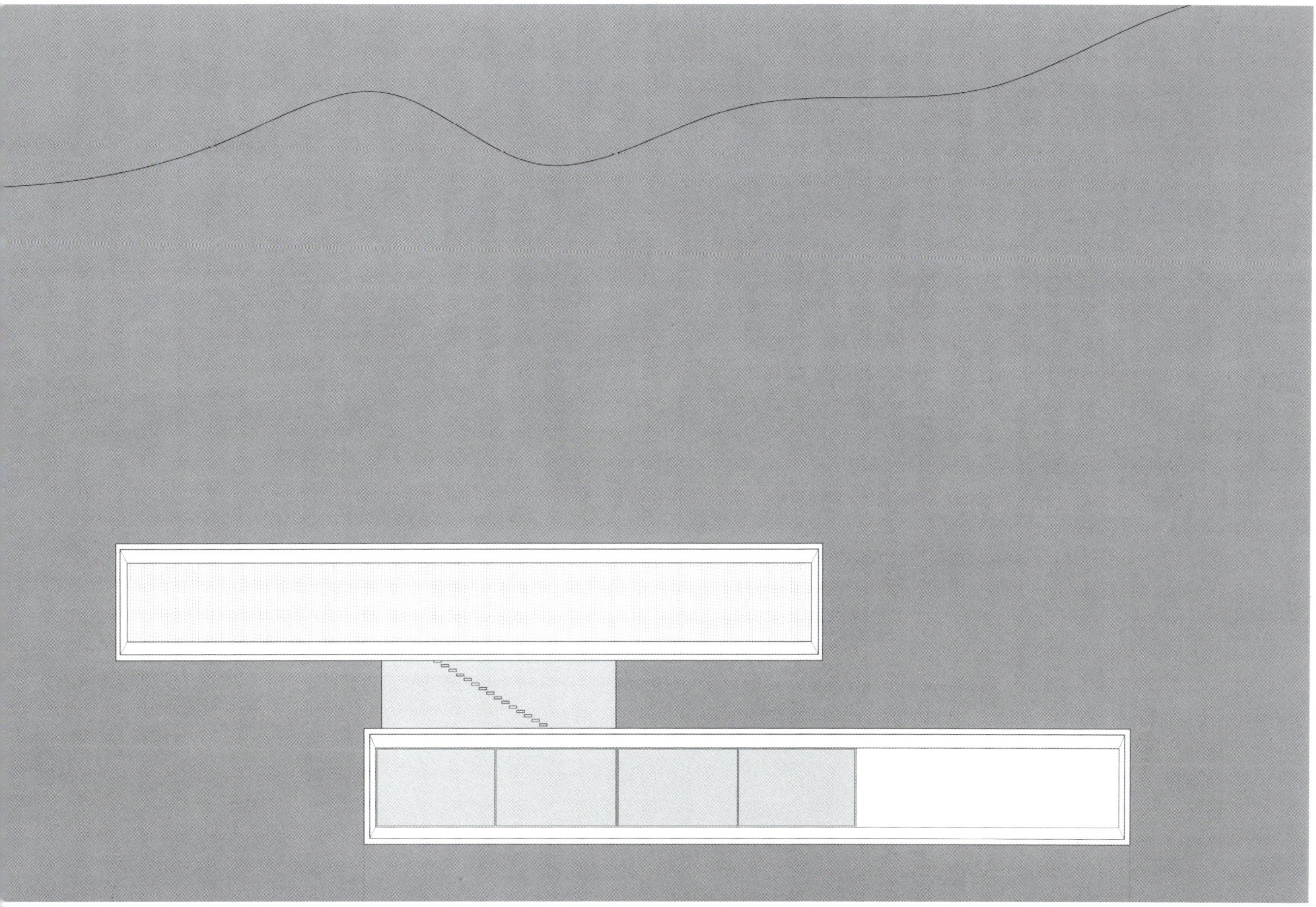

将环境中的海水引入到别墅的露天泳池中，实木的露台一直伸展到水边，无论是在水中畅游，还是在岸边小憩，都能让人充分的融合到自然界中。夜晚，泳池旁边的灯带被打开，随着灯光的合理介入，照亮了天空的同时也照亮了我们的心情。

室内空间中大量的使用，充分证明了就地取材的重要性。粗狂的石材，大气中夹杂着自然。窗帘采用当地的芦苇编制，阳光透过窗帘柔柔的洒进房间，星星点点，错落有致。精选的家具和整体粗放的建筑形成了激烈对比，营造出一种细腻的慢生活。V

Project: House +
Architects: Tomasz
Location: Galowice, Poland
Design: PAG design firm

In Europe, Poland has a good economic foundation is undoubtedly a bright star. Poland to offer designers many other European countries can not give the opportunity to Poland is clearly attracted many famous European design team reasons. Of course, we can not ignore the strength of the Polish local design team, their work to explore and unique vision against all the odds, at the last Venice Biennale was praise, had a great impact. In recent years, with the design of the continuous exploration and innovation, several young Polish architectural design team enjoys growing reputation overseas by Tomasz PAG led the design firm is one of them.

PAG design firm dedicated to architectural design to interior design from a variety of different scales of projects, his recent works is a city located near Wroclaw in private homes, called the House +. The Sudeten Mountains in the project, the base covers an area of large, dense forests separated from its neighbors and the surrounding. Through the building's design, designers attempt to evoke pre-war Poland Lower Silesian region farm style memories. House + design made good use of local traditional elements, and performance with a very modern way to meet the functional needs of modern families. As the excellent performance of this project, PAG design firms and their works in the recent wide attention.

PAG design firm dedicated to architectural design to interior design from a variety of different scales of projects, his recent works is a city located near Wroclaw in private homes, called the House +. The Sudeten Mountains in the project, the base covers an area of large, dense forests separated from its neighbors and the surrounding. Through the building's design, designers attempt to evoke pre-war Poland Lower Silesian region farm style memories. House + design made good use of local traditional elements, and performance with a very modern way to meet the functional needs of modern families. As the excellent performance of this project, PAG design firms and their works in the recent wide attention.

Designer of the building exterior wall treatment is undoubtedly the House + design a highlight. To make building blocks to maximize unity of body, designers will be the roof and outer wall stickers decorated with the same veneer materials - fiber cement tile. This kind of external walls and roof slope to impose the same kind of material relevance to present uniform approach, in a recent Dutch architecture in common. Material compared to the thick texture of building facades, interior designers on the treatment of the surface it appears simple and light - all with white walls rule. Meanwhile, designer furniture, the construction will minimize the impact, while the outside doors and windows leading to become rich interpretation of the surface of the building some simple incision.

Simple and practical program, on forms and materials reflect the good control of the consistent characteristics of PAG design firm. In the House + design, simple wall treatment is undoubtedly the climax of architectural design, streamlining the way the performance of the overall style makes the building to achieve maximum unity. Simplification of the expression, tend to emphasize on building their own values, and his supreme sense of beauty and architecture and left foot in a dialogue space, and one in the true meaning, one can only with Meige typical of the sensitive can appreciate.

House +

项目名称: **House +**
建筑设计: **Tomasz**
项目地址: **Galowice**, 波兰
设计公司: **PAG**设计事务所

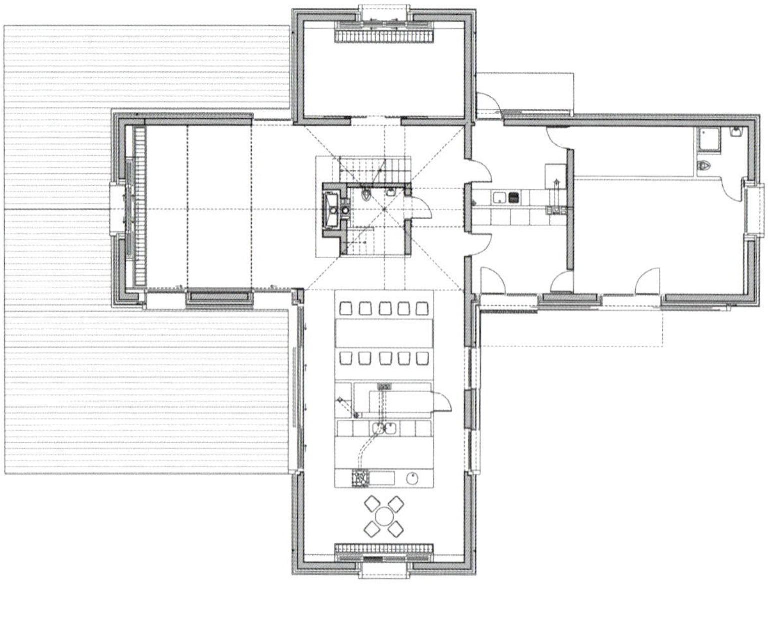

House +的设计很好地运用了当地传统元素，并借助十分现代的表现方式来满足现代家庭对功能的需求。由于在这个项目上的优秀表现，PAG设计事务所及其作品在近期广受关注。这栋覆盖着倾斜屋顶的建筑，由两个向外伸展的体块组成，在平面上形成了一个四翼交叉体。

PAG设计事务所的设计师们常常会针对设计中存在的不同问题，寻找独特而恰当的解决方法。他们总是有区别地对待设计中的每个问题，最终得出独一无二的解决方案。在这个案子中，设计师主张在充分尊重投资者要求和功能需要的前提下，极力强调基地的独特个性。

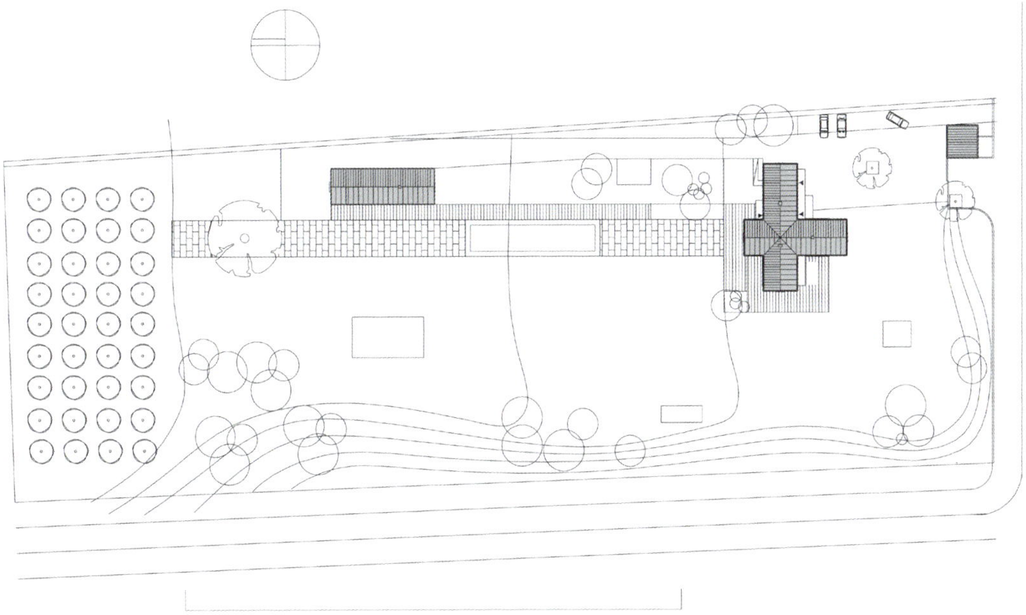

设计师对建筑内外墙面的处理无疑是House +设计的一大亮点。为了使建筑体块能够最大限度地统一，设计师将屋顶和外墙面贴饰同一种贴面材料——纤维水泥瓦。这种对外墙和倾斜的屋顶施以同种材料，使其呈现均一关联性的处理方法，在近期的荷兰建筑中屡见不鲜。相比于建筑外立面材料的厚重质地，设计师们对内墙表面的处理方式则显得轻巧简约——用白色墙壁统治一切。同时，设计师将家具对建筑的影响降到最低，而把通向室外的门窗演绎成为这个拥有丰富表面的建筑的一些简单切口。在House +的设计中，简洁的墙面处理无疑是建筑设计的高潮所在，精简的表现方式使得建筑在整体风格上达到最大限度的统一。V

Project: House in Sakuragawa
Architects: Makoto Tanijiri
Design team: Hayato Komatsu(ex-staff), Yoshio Ohno(ex-staff)
Location: Tokyo, Japan
Building area: 91.01 sqm
Photographer: Toshiyuki Yano / Nacasa&Partners Inc.

This is a real little house by the famous Japanese Suppose Design Office completed the design studio, located in Tokyo Itabashi a T-junction, an area of only 50m^2, but also the total living area 91m^2. Rather than the rule in such a small area on the interpretation of a family home to a warm place to live so vividly, and only grows in the densely populated Japanese architect can. Designed to overcome the geographic and subject area in question, based on the limited space molded into a bright and spacious indoor environment for human habitation.

Architects to adopt a hierarchical layout approach, and to minimize the division of interior space and visual barriers, so that each layer and between the upper and lower floors seem transparent flow of invisible in the visual sense to expand the interior space. Ceiling and floor tiles are used the more light-colored wood, furniture, colors are generally light elegance. White walls and white metal staircase natural connection for each floor, from floor to floor and then to the top balcony, smooth writing at one go. It is no exaggeration to say that in such a house, a parent with a child, elderly people, and even then a dog spacious enough. Integrated space design allows families to communicate at all times, in any corner of the house can feel each other's existence, the family filled with happiness everywhere.

Building is divided into 3 layers. The first floor is a side street half-closed garage and the relative privacy of the bedroom space. On the second floor is the family living room and the main restaurant. South of the house is a park, so a fan and open windows, of course, the South is the best choice.

Sitting on the sofa reading the newspaper in the afternoon or early morning cleaning the busy packing, leisure or spend more looked up looked, the children playing in a park not far away, across in front of the landscape throughout the year, but also how warm and cozy. Long glass windows to the outdoor scenery are diverted to the chamber, not the cramped and bound, under a beautiful little house containing four seasons. Through the third floor roof terrace, glass curtain wall to increase the brightness of indoor and transparent sense, once again, the overall design theme.

Whether the building facade or interior design style, furniture selection, all continue Suppose Design Office studio has always been light and breezy style. In view of topographic position, the external walls of buildings designed to be more conceptual, wood mosaic, but the choice of steady low-key color for a home construction, fashion, hidden under the reserved style, highlighting rather than obvious, modern but not cold.

Small and cozy, living in this house, and slowly enjoy the four seasons, enjoy taste of life.

House I

sakuragawa

项目名称：**House in Sakuragawa**
建筑设计：**Makoto Tanijiri**
设计团队：**Hayato Komatsu(ex-staff), Yoshio Ohno(ex-staff)**
项目地址：日本东京
建筑面积：**91.01** 平方米
图片：**Toshiyuki Yano / Nacasa&Partners Inc.**

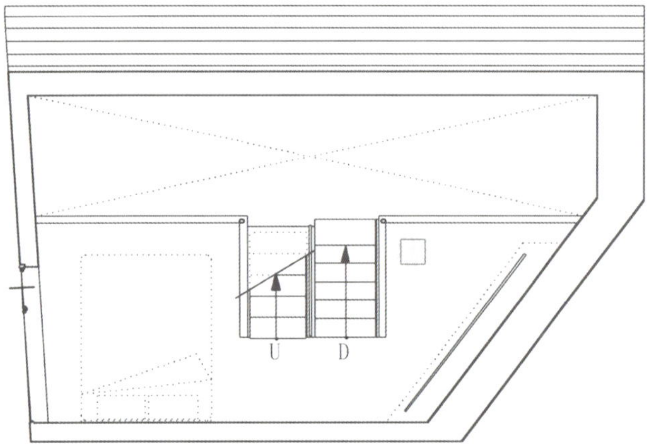

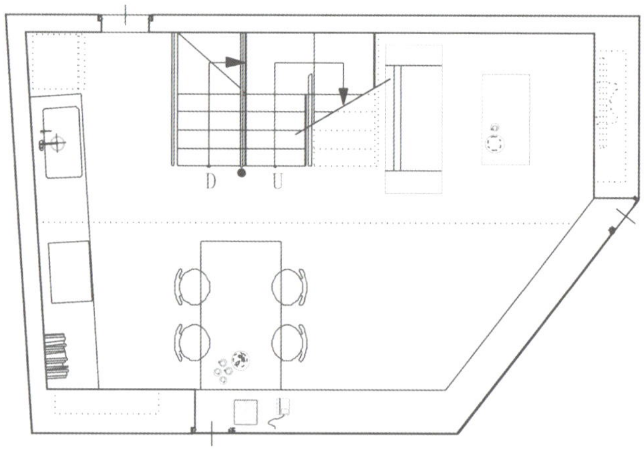

这是一座名副其实的小房子，由日本著名的Suppose Design Office工作室设计完成，位于东京板桥的一个丁字路口，占地面积只有50m²，总居住面积也不过91m²。在如此狭小而不规则的面积上把一座家庭住宅演绎地如此生动温馨适合居住，也只有生长在人多地少的日本建筑师才能做到。设计的主旨是在克服地理位置和面积问题的基础上，把有限的空间塑造成为明亮宽敞适宜人居的室内环境。建筑师采用了分层布局的方式，并尽量减少室内空间的分割和视觉障碍，使每一层及上下楼层之间显得通透流通，无形之中在视觉上扩大了室内的空间感。

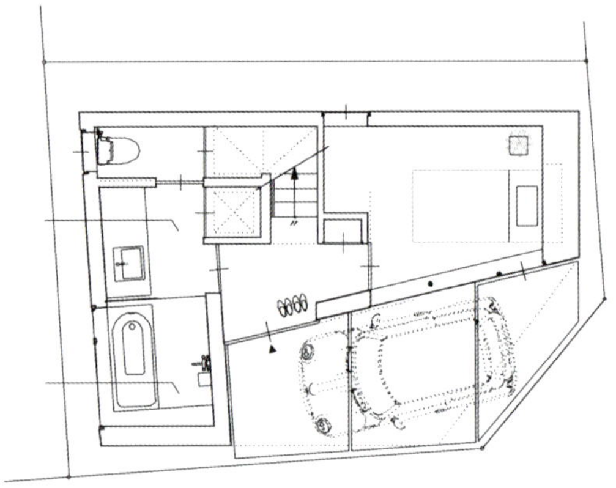

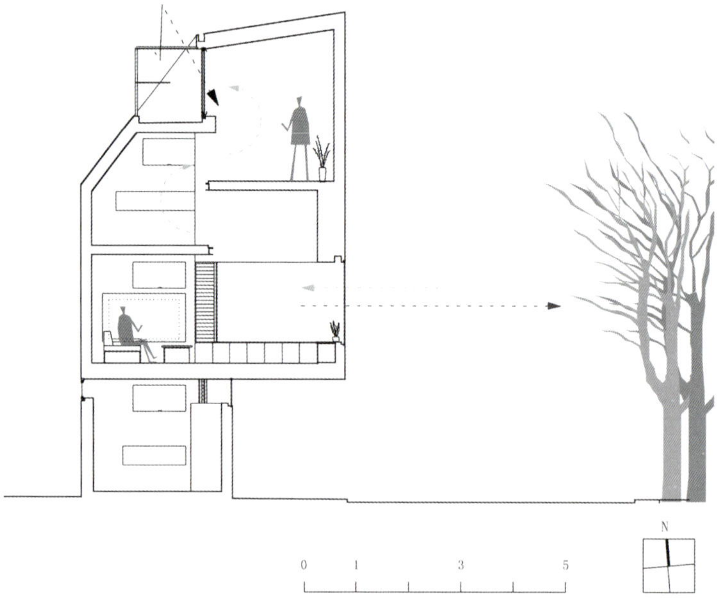

天花和地板都采用较为浅色的木纹，家具的颜色也普遍清淡素雅。白色的墙面和白色金属楼梯自然连接每个楼层，从一楼至三楼再到顶层露台，行云流水一气呵成。可以毫不夸张地说，在这样的房子里，一对父母加上一个孩子，还有老人，甚至再养条狗也足够宽敞舒适。融为一体的空间设计使家人时时刻刻可以交流沟通，在房子的任何一个角落都可以感觉到彼此的存在，家庭的幸福四处洋溢。

建筑分为3层。一楼主要是面街半封闭式的车库和相对私密的卧室空间。二楼主要是家人的起居室和餐厅。房子的南面是一个公园，所以一扇面南而开的窗户当然是最好的选择。坐在沙发上看书读报的午后，或者是忙碌收拾打扫的清晨，闲暇或者忙碌之余抬眼望去，孩子在不远处的公园玩耍，一年四季的景色在眼前划过，又是何等的温暖惬意。玻璃长窗将室外的风景都引到房间里来，没有了局促和束缚，小小的房子盛下了美丽的四季。三楼直通楼顶露台，玻璃幕墙增加了室内的明亮度和通透感，再次体现整体的设计主旨。

鉴于地形位置，建筑的外墙设计得比较概念化，原木拼接，但是颜色的选择稳重低调，对于一个家居建筑，时尚下隐藏含蓄内敛，突出而不张扬，现代而不冷漠。小而温馨，生活在这样的房子里，慢慢享受四季，尽情品味生活。V

Project: The Letter Box House in Blairgowrie
Architects: Rob McBride Debbie-Lyn Ryan
Location: Blairgowrie, Australia
Building area: 290 sqm
Photographer: John Gollings

Sea View Villa is located in a neighborhood, architects and the hollowing out of the way through the extrusion to the disposal of construction and the relationship between the surrounding environment. Looked from one end of the building, a house by a small aperture to the horizontal and vertical gradually expanded to include a black metal roof and roof eaves, can be regarded as a continuous surface. This building villas, such as a camel sculpture is placed in an artificial desert, and the street on both sides of the typical eclectic style traditional villa stands in sharp contrast, very impressive. In the overall layout, the architects used the number "7" shape, unique at the same time corresponds with the building of streets. And how to make the shape of "7" of the irregular geometry to adapt to the environment, is thought worthy of the architect. Typically, according to local architecture and site planning is not allowed too close to the border, need to allow enough distance to shy away.

However, both the building itself, "mailbox" in the role of public facilities, so its amount and form of the entire body must be part of the barrier to be out of the wall outside. So, for the site boundary relations of uncertainty, the architect used the fuzzy boundaries between the fence and the way the site. Meanwhile, the external facade, a simple unified package of materials emphasize the effect of the external volume, while the elevation of the house also made of wood materials, the construction of body hollowed out to do the ultimate sense. The second floor balcony of the deal, the architect through the package approach to the house facade outer surface, so that the outer surface of large-scale use of wood materials from the white color and contrast decorative balcony, two kinds of materials in the sensory reached conflicting visual effects. This experiment in diversity of wood used in modern architecture possible, also appeared in other Australian architects design, such as Godsell and so on.

As one has five spacious bedroom holiday cottage, even many members of the family can easily stay. Can also have a recreational use for the open meeting room. The highlight of the entire house design interior and exterior wall shown to reverse the form with folded echo. Behind the wall south elevation is a similar service space corridor connecting the two front bedrooms. The service space can be interpreted as the functional areas of a partition, make space more orderly. Hidden within the steel members, the red wooden box beam structures such as buildings designed in the architect to the next visible.

The Let

er Box House
n Blairgowrie

项目名称: **The Letter Box House in Blairgowrie**
建筑设计: **Rob McBride Debbie-Lyn Ryan**
项目地址: 布莱尔高里, 澳大利亚
建筑面积: **290** 平方米
图片: **John Gollings**

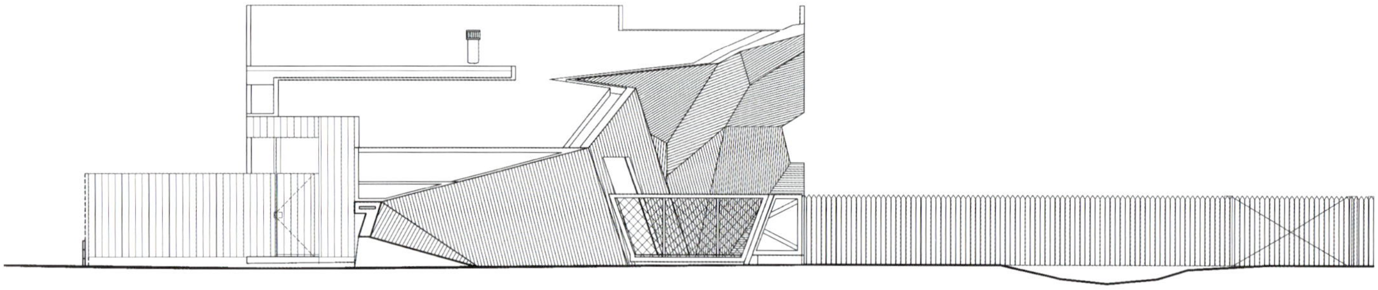

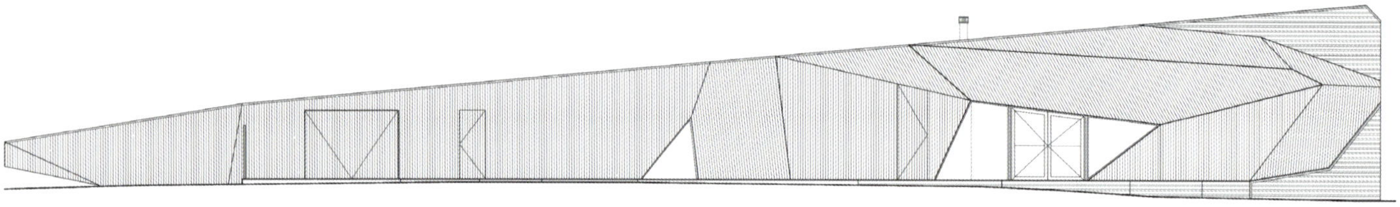

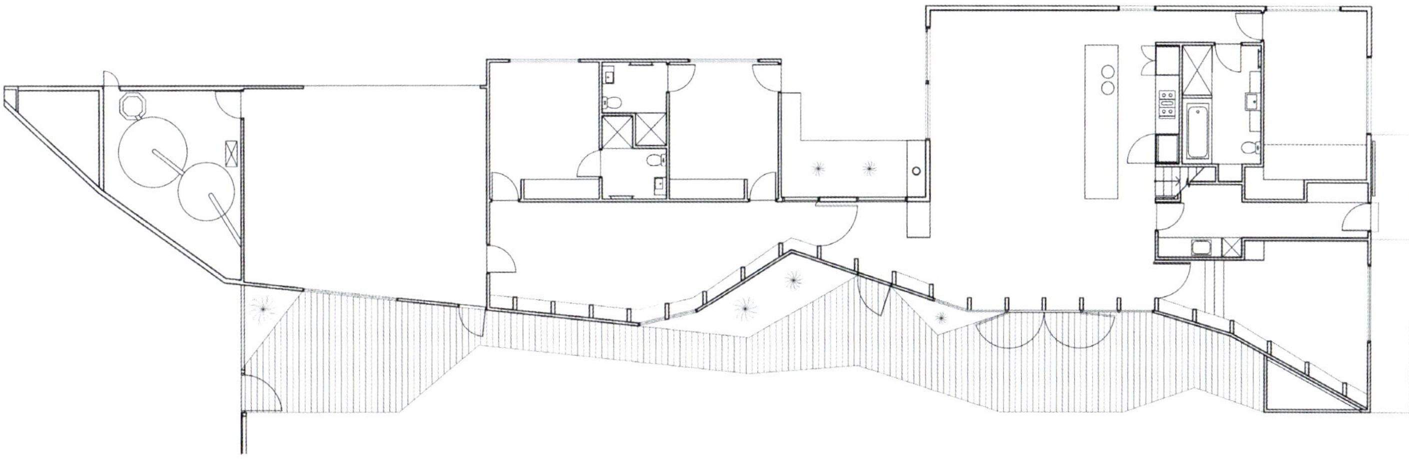

别墅位于一个海景街区之中，建筑师通过挤压和挖空的手法，来处理建筑与周围环境之间的关系。从建筑的一端望去，房子由小小的信箱口向横向和纵向慢慢扩大，包括屋顶和屋顶黑色的金属屋檐，都可以看作是一个连续的表皮。这幢别墅如雕塑般地被放置在一个人造沙漠中，与街道两边典型的折中主义式传统别墅形成了鲜明对比，十分引人注目。在总体平面布局上，建筑师采用了数字“7”的形状，造型独特，同时又与街道的门牌号相对应。而如何让这个形似“7”的不规则几何体适应周围环境，是值得建筑师思考的。通常情况下，根据当地规划是不允许建筑与场地边界太过接近，需要留出足够的退让距离。

但该建筑本身兼有“邮箱”的公共设施作用，因此它的整个体量与形式必须有部分要出于围墙护栏之外。于是，针对场地边界关系的不明确，建筑师采用了模糊围栏和场地之间界限的手法。同时，在外墙立面上，简单统一的材料强调外部体积包裹的效果，而房子内立面亦采用木质材料，将建筑体量的挖空感做到了极致。在二楼阳台的处理上，建筑师通过包裹房子立面的方法处理外表面，让外表面大量使用的木质色彩与由白色装饰的阳台形成鲜明对比，两种材料在感官上达到了相互对立的视觉效果。这种将木材多样地使用在现代建筑中，也出现在其他澳大利亚建筑师的设计中，如肖恩•高德尔森等。

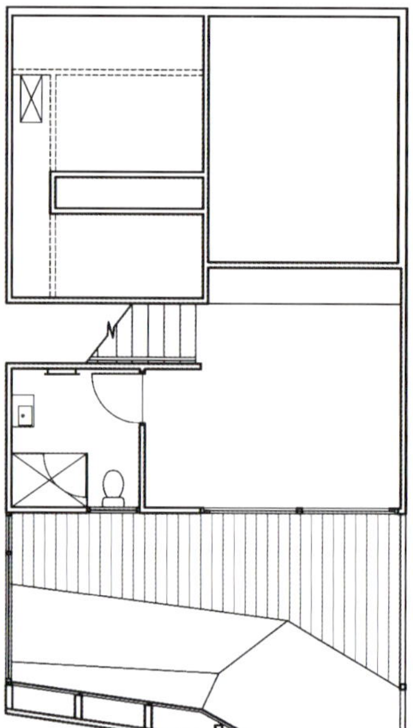

作为一幢拥有五间宽敞卧室的度假小屋，即便是有很多成员的大家庭也能轻松入住。同时还拥有一个可以供聚会休闲使用的开放式客厅。整个房子设计的亮点是室内表现出来的与外部墙体扭转和折叠形态的呼应。南立面外墙背后是一个类似走廊的服务空间，连接前部的两间卧室。这一服务空间可以被理解为将功能区进行了一个分割，使空间更加有序。内部隐藏的钢构件、红色木质箱型梁等建筑结构都在建筑师的精心设计下清晰可见。V

Project: Wa Sun Jai
Architect: HyoMan Kim - IROJE KHM Architects
Design team: SuMi Jung, MiYeong Park
Structure designer: Guang Min Jeon
Location: GungNaeDong, BunDangGu, SeongNam, GyeongGiDo, Korea
Building area: 515 sqm
Photographer: JongOh Kim

The owner of house is born in Korean-style house and grows up in this land, as well as, the village is a place with ancestral burial ground. The place is the origin of life for him and a place where the trace and remembrance of life is compressed. Now he, who composed a household of family, came back to the land and build up the land of life for three generations with his mother who kept the site until now.

A pond snail lays young in its' skin in the peel and the young grows up eating the skin of mother. If the young all grows up, the mother left only with 'peel'. It is an instinctive devotion to inherit the life. Also, it is a 'virtue' that is a destiny of life. Maybe he tried to realize the virtue of pond snail in the site where ancestors' devote is piled. It is tried to build a 'layer' only as a 'peel'.

An architectural 'wall' only as a 'peel' is built and the wall endowed with a role to mean an idea of time. The 'oldness' of Cor-Ten Steel is an instinctive peel. The wall and the wall of exposed concrete are an 'inner skin' of past story as well as a peel of sometime in the past. The walls with symbolic time, divide the site into 'garden' of ten 'places' the site and each garden, as a land where records the history will latent another program of future life. And the connection of garden and the story of garden are expected to show plentiful spatial drama and story of life.

With the existing Korean-style house as the central figure, the composing element like outer wall, inside fence, outside fence of every space built a 'wall' that only has simple 'side' as a 'form' With this, the Korean-style house is symbolically embossed from every direction, and become an object. It will function as 'landscape of memory' to the residence and the visitors of the house.

The urban landscape of high-storied office building in Bundang new town with background of tiled roof on the Korean-style house that looks down from the second floor makes to feel the mystery dissonant harmony and difference of periodical character of culture. Moreover, the firmness of cumulated time may be a reason that the grandeur of the small tiled house could see.

The new residential mass is long arranged surrounding the Korean-style house. With this, the mass is being a fence of outside. Moreover, it could form an outside space between the existing Korean-style houses.

wa sun jai

项目名称: **Wa Sun Jai**
建筑设计: **HyoMan Kim - IROJE KHM Architects**
设计团队: **SuMi Jung, MiYeong Park**
结构设计: **Guang Min Jeon**
项目地址: **GungNaeDong, BunDangGu, SeongNam, GyeongGiDo, Korea**
建筑面积: **515** 平方米
图片: **JongOh Kim**

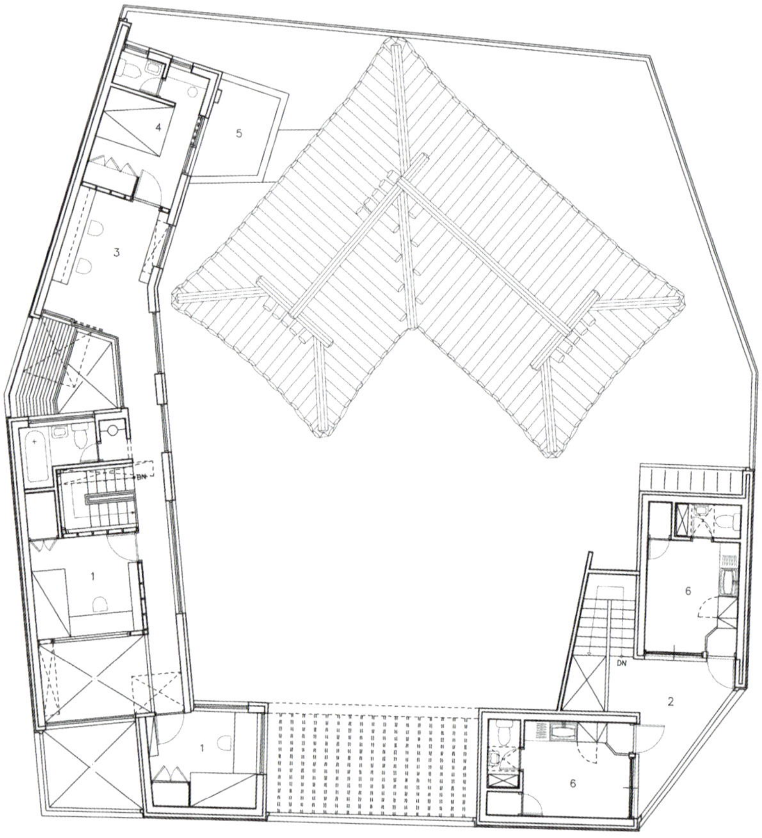

房子的主人生长于韩国风格的建筑环境中。这里是该村一个祖先的墓地，这个地方是他生活和生长的地方，这里的回忆被压缩起来。现在，在这片土地上，建立起世代生活的院落。池塘中，一个幼小的蜗牛在长大的过程中，汲取了母亲的营养。在成长的过程中，母亲给与了很多，这是一种本能的奉献和生命的继承。这样的品德，是生活的命运。也许他试图实现祖先堆积的池塘蜗牛的美德。一个建筑“墙”只能作为“皮”，是建立和与角色赋予了墙壁，意味着时间的观念。墙，清水混凝土墙是内心的外在表现，对过去的故事以及在过去的某个时候的剥离。时间的墙壁上，记录着这片土地的历史，将潜在的另一种未来生活规划。

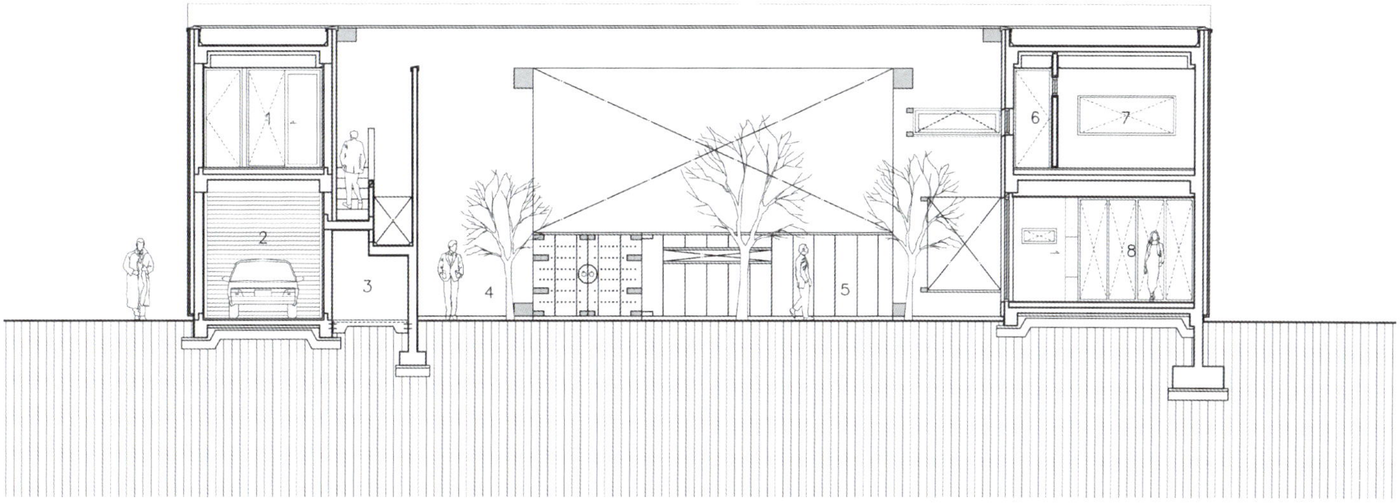

花园和园林的故事，将显示丰富的戏剧和生活空间的故事。 现在韩国人的核心文化，像外墙构成要素的围栏，房子围栏外的每一个空间建成一个“形式”，有了这一个“墙”，具有简单的“面”，浮雕风格的房子则象征着各个方向，这里将成为景观的一个居住地。

采用高层办公楼，屋顶瓦片与韩国风格的房子，在房子上俯视新城镇景观，区分了不和谐、神秘感以及文化的差异。新的住宅周围早已建造了朝鲜式的房子，此外，它可以形成一个与现有的韩式房屋相连的外部空间 。V

Project: Mirindiba's House
Architects: Marcio Kogan
Location: Sao Paulo, Brazil
Building area: 1070 sqm
Photographer: Nelson Kon

Mirindiba's House is one such from the beginning to the end of each step for a perfect example, from design to construction, every detail can be called perfect. This building beautiful villas from Sao Paulo city, located in the city the best residential areas of law and order. In this project, each point of import and export are well thought of. It outlines the construction process from blueprints to the construction of the process is clean and highly organized nature. The construction process so fine, so this building as if from the hand of the artist's process works. It is the solemn atmosphere, with a strong sense of form and extraordinary visual experience, like arts and crafts in the industrial age generally amazing.

This is a long-span structures with wonderful architecture, it has a large living room is almost completely open, the use of large windows, the addition of indoor air permeability, while promoting the exchange and circulation, to free fun at people's feelings. It uses a black Brazilian hardwood, concrete structures and local stone materials in such a frame structure and materials used for building a more rich with meaning.

Mirir

diba's House

项目名称: **Mirindiba's House**
建筑设计: **Marcio Kogan**
项目地址: 巴西圣保罗
建筑面积: **1070** 平方米
图片: **Nelson Kon**

Mirindiba's House就是这样一个从始到终，每一个步骤都进行的堪称完美的范例，从设计到施工，每一个细节都可以称得上天衣无缝。这栋漂亮的别墅远离圣保罗市区，坐落地点是这个城市治安最好的住宅区。在这个项目中，每一个出入口的角度都是经过深思熟虑的。它的建筑过程，从蓝图的勾画到施工的过程都是干净而富有组织性的。如此精细的建造过程，让这栋建筑犹如出自艺术家之手的工艺作品。它端庄大气、拥有强烈的形式感和非同凡响的视觉体验，就像是出现在工业化时代的手工艺品一般让人惊叹。这是一个有着精彩大跨度结构的建筑，它拥有一个很大的几乎是完全开放的起居室，大面积落地窗的运用，在增加了室内的通透性的同时促进了空气的交换和流通，给人自由畅快的感受。它运用了黑色的巴西硬木、混凝土结构和本地取材的石材，这样的框架结构和材料搭配选用让建筑更加富有内涵。在炎热的夏季，楼下开放的泳池中，人们可以畅快享受戏水的乐趣，累了还可以在游泳池边的阴凉处享受悠闲的午后时光。

Project: So Da Hun
Architect: HyoMan Kim - IROJE KHM Architects
Design team: Hyun Jung Ju
Structure designer: MOA. Sung Yeong Oh
Location: 229-2 Bigeun-dong, Pyeongtak, Gyeounggi-do, Korea
Building area: 63.25 sqm
Photographer: Jong Oh Kim

This site that is located in the heart of downtown is a small, irregular, polygonal shaped land and it is ragged old space, only area of 100 sqm. At the same time, the maximum building area is no more than 60 sqm.

While adapting the skipfloor system, it was possible to maximize the visual range and to produce the interesting space of variety, as possible as this small site can do.

Because of the boundary of small courtyard and living room and the window between the concrete fence and interior space are transparent, the extended outdoor garden space and interior space maximize the horizontal visual range.

The translucent images which are consisted with stainless mesh that filtrate the sight from opposite church and side stores and direct sunlight, are the major impression of this house.

Small courtyard was covered with wood panel, for the maximum efficiency of use of the place. This spacs of inner courtyard where is surrounded with stainless mesh and exposed concrete walls of this house, is semi interior space.

From ground courtyard level to roof garden level, several trees and flowers were planted vertically, on the canopy of main gate and on the balcony of upper floor and on the roof garden floor.

The name of this house "So Da Hun" means "small house but the house of rich space". This project was a research about what is the rich spatial system in the minimum area of site.

So Da Hun

项目名称: **So Da Hun**
建筑设计: **HyoMan Kim - IROJE KHM Architects**
设计团队: **Hyun Jung Ju**
结构设计师: **Sung Yeong Oh**
项目地址: **229-2 Bigeun-dong, Pyeongtak, Gyeounggi-do, Korea**
建筑面积: **63.25** 平方米
图片: **Jong Oh Kim**

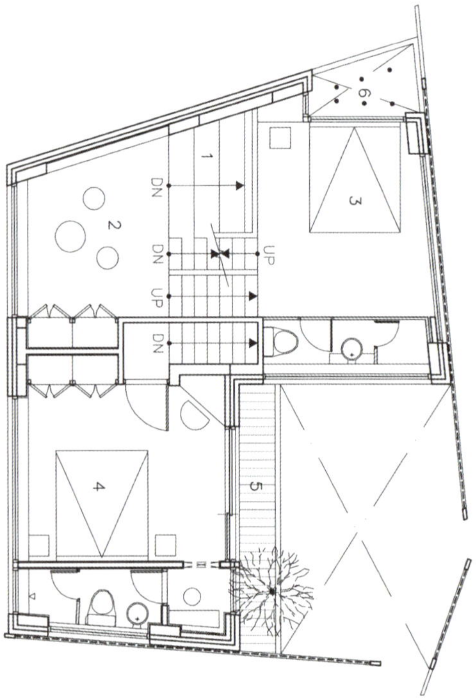

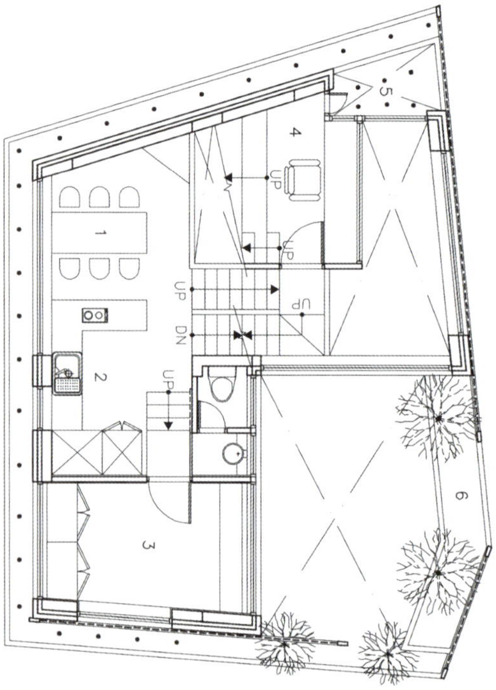

这幢金属网建筑，位于市中心，一个小的，不规则的，多边形形状的土地，这个破败的旧空间，只有100m²。与此同时，最大建筑面积不超过60m²。同时适应了skipfloor系统，有可能最大限度地拓展视觉的范围，并生产出各种有趣的空间，尽可能将这个设计实施。

由于小庭院和客厅之间的混凝土围墙和内部空间窗口边界是透明的，因此延长室外园林空间和室内空间，视觉上最大限度地提高美学水准。半透明泛光效果，是通过不锈钢网组成的过滤形成的，从对面的商店、教堂和侧面图像阳光直射进来，形成了这所房子的主要印象。

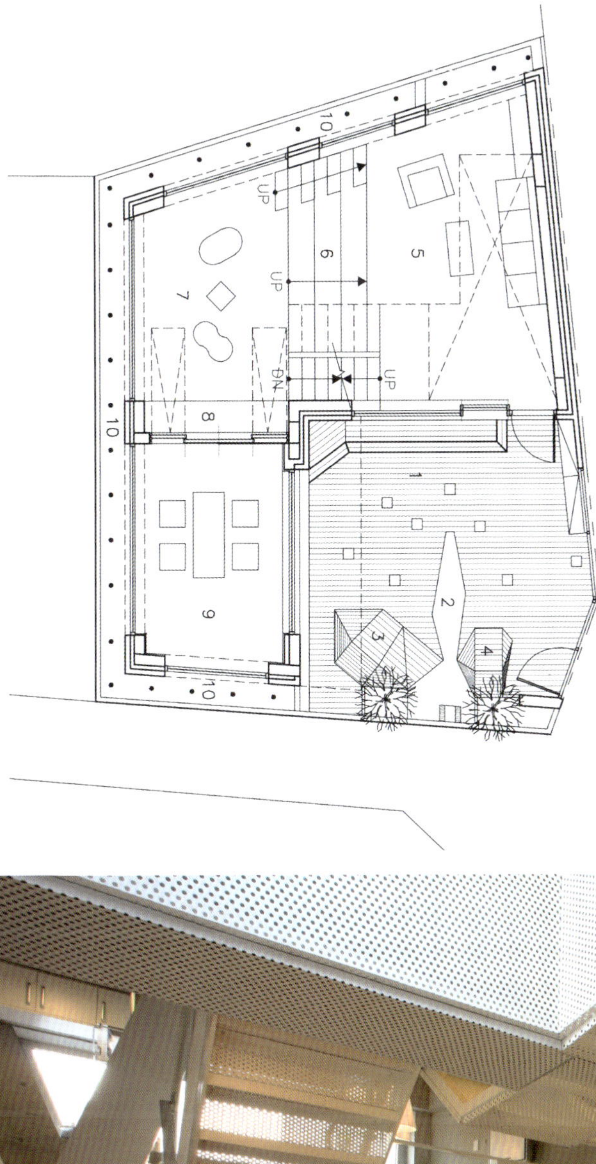
UP
UP
DN
UP
1
2
3
4
5
6
7
8
9
10
11

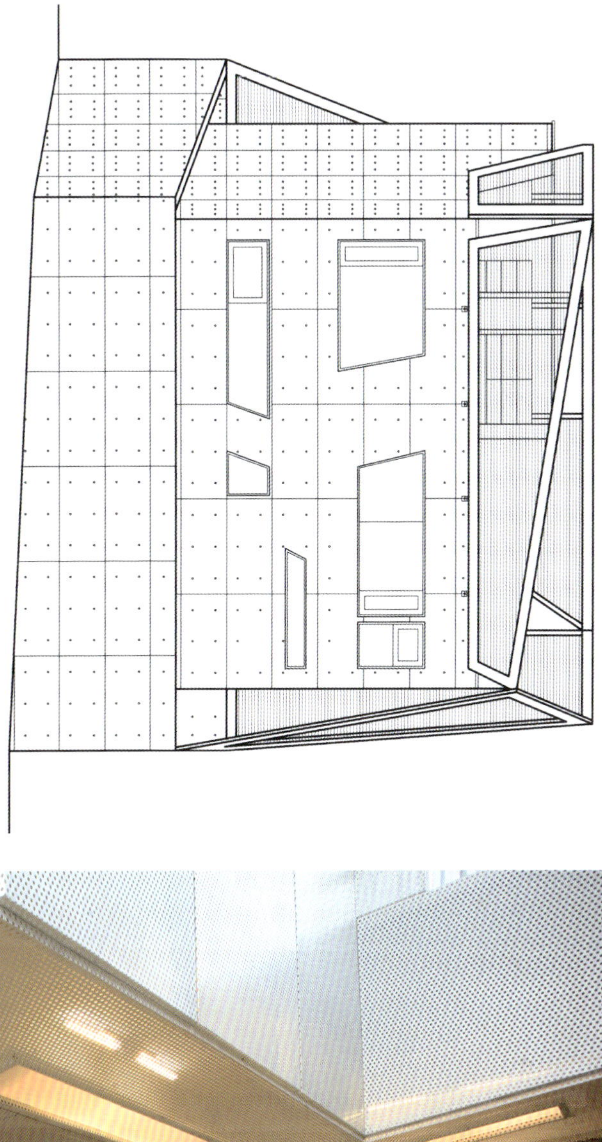

小院上覆盖着木板，为的是提高场所使用效率。庭院在四周用上不锈钢网，暴露了这所房子的水泥墙，也是个半室内空间。从地面到屋顶的花园庭院级水平，加之一些树木和花卉种植垂直布置，在大门上的华盖、楼上阳台和屋顶的花园绽放。这个房子名字的意思是“小房子，但是个丰富的家”。这个项目是有关最小面积空间系统的研究。V

Project: Corten house
Architects: Marcio Kogan
Location: Sao Paulo, Brazil
Photographer: Nelson Kon

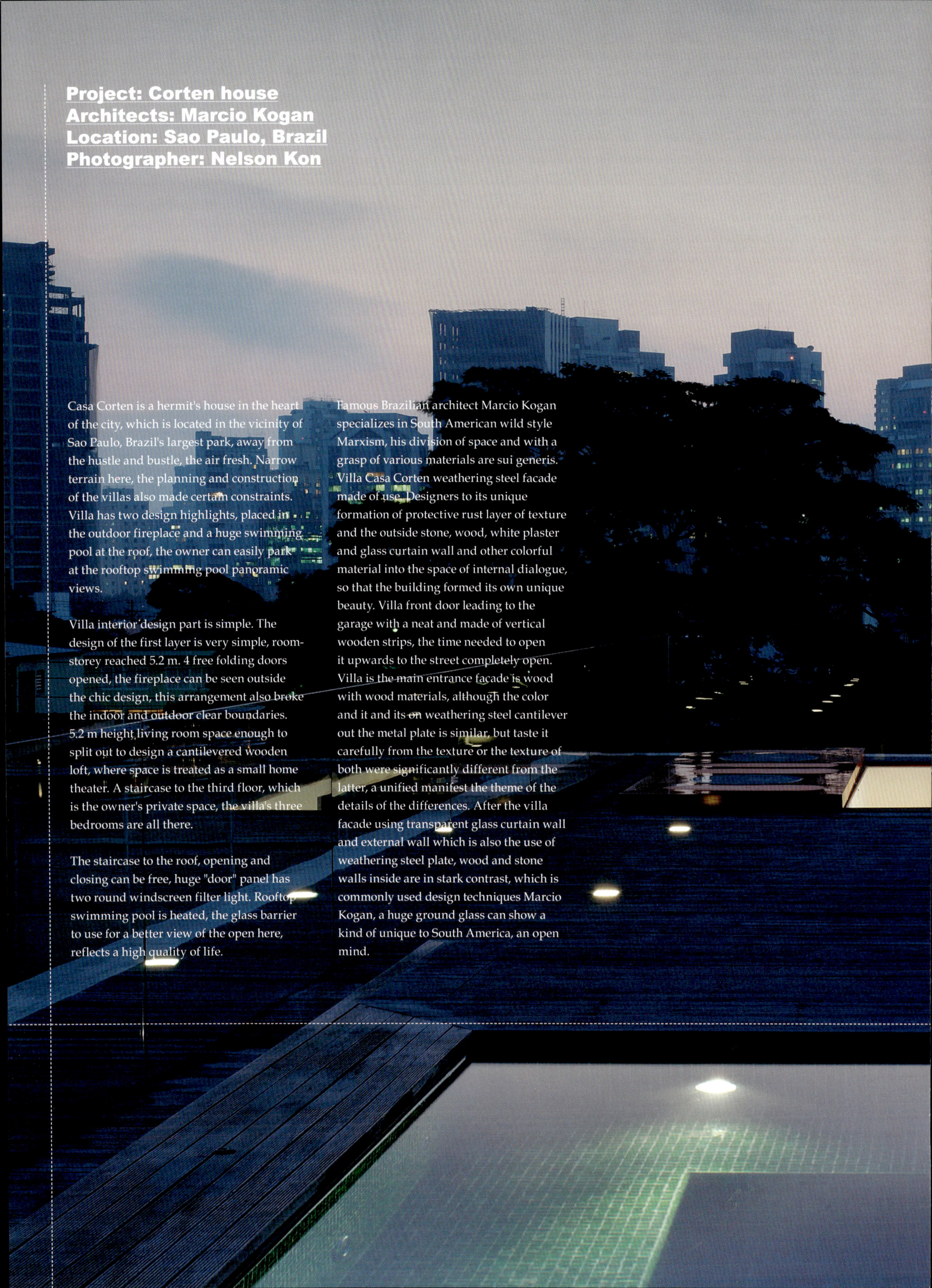

Casa Corten is a hermit's house in the heart of the city, which is located in the vicinity of Sao Paulo, Brazil's largest park, away from the hustle and bustle, the air fresh. Narrow terrain here, the planning and construction of the villas also made certain constraints. Villa has two design highlights, placed in the outdoor fireplace and a huge swimming pool at the roof, the owner can easily park at the rooftop swimming pool panoramic views.

Villa interior design part is simple. The design of the first layer is very simple, room-storey reached 5.2 m. 4 free folding doors opened, the fireplace can be seen outside the chic design, this arrangement also broke the indoor and outdoor clear boundaries. 5.2 m height living room space enough to split out to design a cantilevered wooden loft, where space is treated as a small home theater. A staircase to the third floor, which is the owner's private space, the villa's three bedrooms are all there.

The staircase to the roof, opening and closing can be free, huge "door" panel has two round windscreen filter light. Rooftop swimming pool is heated, the glass barrier to use for a better view of the open here, reflects a high quality of life.

Famous Brazilian architect Marcio Kogan specializes in South American wild style Marxism, his division of space and with a grasp of various materials are sui generis. Villa Casa Corten weathering steel facade made of use. Designers to its unique formation of protective rust layer of texture and the outside stone, wood, white plaster and glass curtain wall and other colorful material into the space of internal dialogue, so that the building formed its own unique beauty. Villa front door leading to the garage with a neat and made of vertical wooden strips, the time needed to open it upwards to the street completely open. Villa is the main entrance facade is wood with wood materials, although the color and it and its on weathering steel cantilever out the metal plate is similar, but taste it carefully from the texture or the texture of both were significantly different from the latter, a unified manifest the theme of the details of the differences. After the villa facade using transparent glass curtain wall and external wall which is also the use of weathering steel plate, wood and stone walls inside are in stark contrast, which is commonly used design techniques Marcio Kogan, a huge ground glass can show a kind of unique to South America, an open mind.

casa corten

项目名称: **Corten house**
建筑设计: **Marcio Kogan**
项目地址: 巴西圣保罗
图片: **Nelson Kon**

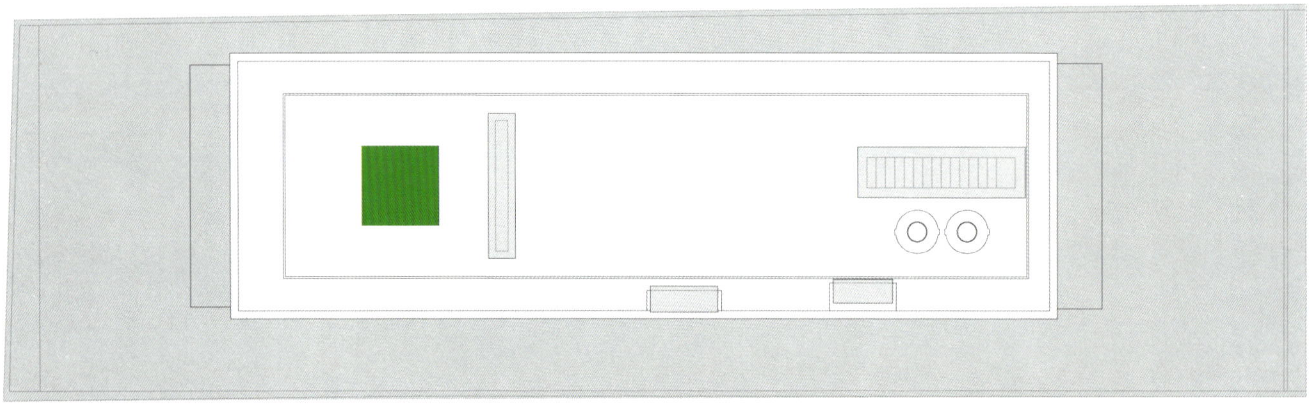

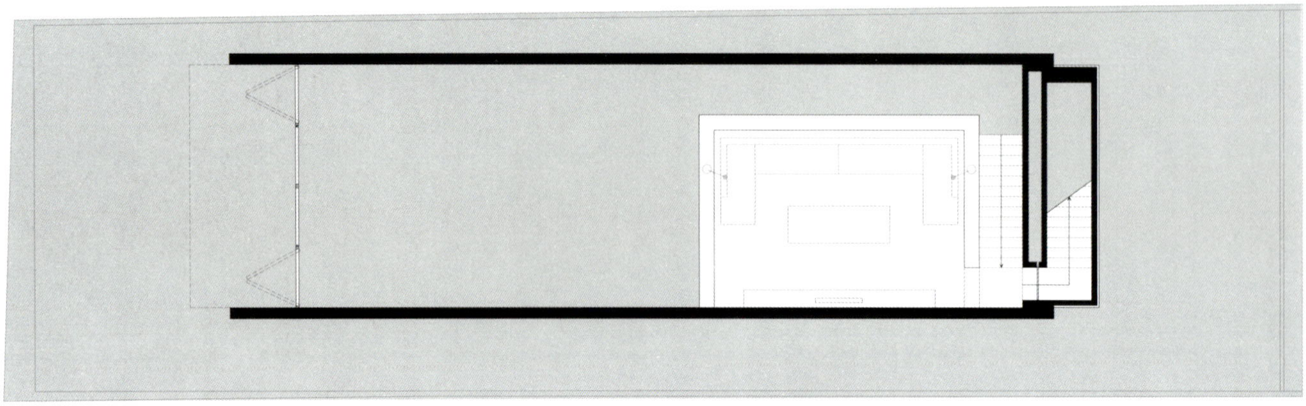

Casa Corten是一座隐逸在繁华都市中的别墅，它坐落于巴西圣保罗最大的公园附近，远离喧嚣，空气清新。这里的地势狭长，对别墅的规划和建造也提出一定的限制要求。别墅有两个设计亮点，放置在室外的壁炉和位于屋顶的巨大游泳池，主人在屋顶游泳池可以轻易地将公园全景尽收眼底。

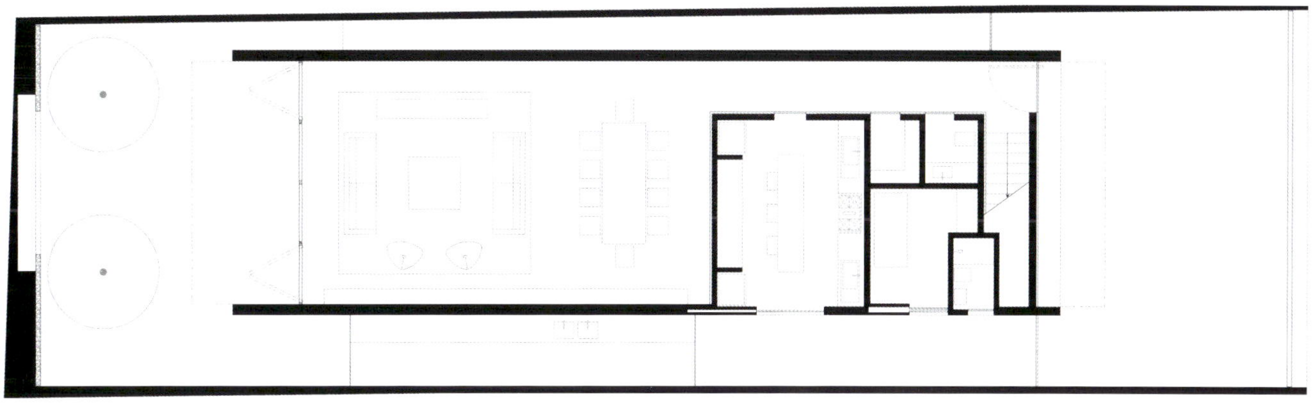

别墅的室内设计部分相当简洁。第一层的设计很简单，房间的层高达到了5.2m。四扇自由折叠门打开后可以看到外部壁炉的别致设计，这样的安排同时也打破了室内外分明的界限。起居室内 5.2m的高度也足够将空间分割，设计出一个悬挑出来的木质阁楼，这里的空间被当作一个小型的家庭影院。一个楼梯通到三楼，那是主人的私密空间，别墅的三间卧室都在那里。通往天台的楼梯，可以自由开阖，巨大的“门”板上设有两个圆形挡风玻璃过滤光线。屋顶游泳池是恒温的，玻璃护栏的使用让这里的视野更加的开阔，体现了一种高品质的生活。

巴西著名设计师Marcio Kogan擅长南美野性主义设计风格，他对空间的分割和对各种材质的搭配把握都自成一格。Casa Corten别墅立面使用耐候钢制作而成。设计师将其特有的保护性锈层形成的肌理与外部的石材、木材、白色灰泥以及玻璃幕墙等各色材质建成的空间进行内在的对话，使整个建筑形成了自己别致的美感。通向别墅车库的前门用整齐且垂直的长条木板制成，需要的时候它可以朝上开启向街道完全开敞。

Project: Circa on Jellicoe
Architect: Pierre Swanepoel
Location: Rosebank, Johannesburg, South Africa
Photographer: Tristan McLaren & studioMAS

The architectural form has grown from the constraints of a narrow site and has resulted in a sculptural landmark. Fins and scrims create visual linkages into and out of the building, while the main gallery remains cocooned from the hustle and bustle.
The design is based on an elliptical plan that does not represent an unsurprising constancy, but rather something of more organic origins. It reflects the abstract nature of art, stimulating thoughts from which the building is to be understood. As its name suggests, it is not specific or defined, it is Circa.

The main architectural features are anodised aluminium fins that partition inside and outside space forming a second skin; a scrim. Their repetitive placement along the facade creates a monumental sculptural form. As elements, their power lies in the play of light and shadow along the edges of the building. Evolving, with the time of day and seasons, the experience of filtered light in the northern edges of Circa accentuates the sensory discovery of space. Playfully poetic, the fins are equal neither in length nor in colour. The inspiration for this came from nature, with its paradox of implied order and chaos. As with the vertical elegance of reeds and grasses in traditional Zulu Kraals, these elements create enclosure while maintaining views through them.

The fire escape is enclosed within a steel structure that is separate to the main building. Covered by wire mesh so that creepers can grow to cover it, this passage forms unique point of observation of the stateliness of the fins and the ellipse. This dialog between handmade product (the facade) and natural growth (the green scrim) aims to evoke an appreciation of the poetic and functional importance of scrims in our cities.

Circa comprises three floors. The ground floor is named Speke and consists of about 106m^2 of exhibition space for crafts. On crossing the public threshold, users are drawn into the building via a perimeter ramp that connects ground and first floor exhibition spaces and promotes view through the fins into the city beyond.
Its double volume first floor, consists of 177.76m^2 multi-purpose exhibition space, with 7 movable display screens which can be dropped through the floor into the ground level below, realising the full extent of the multi-purpose floor. The top floor consists of the Darwin Room, an 85m^2 private lounge that can be rented out for functions and that spills out onto a 20m^2 deck, overlooking Johannesburg.

on Jellicoe

项目名称: **Circa on Jellicoe**
建筑设计: **Pierre Swanepoel**
项目地址: 南非约翰内斯堡
图片: **Tristan McLaren & studioMAS**

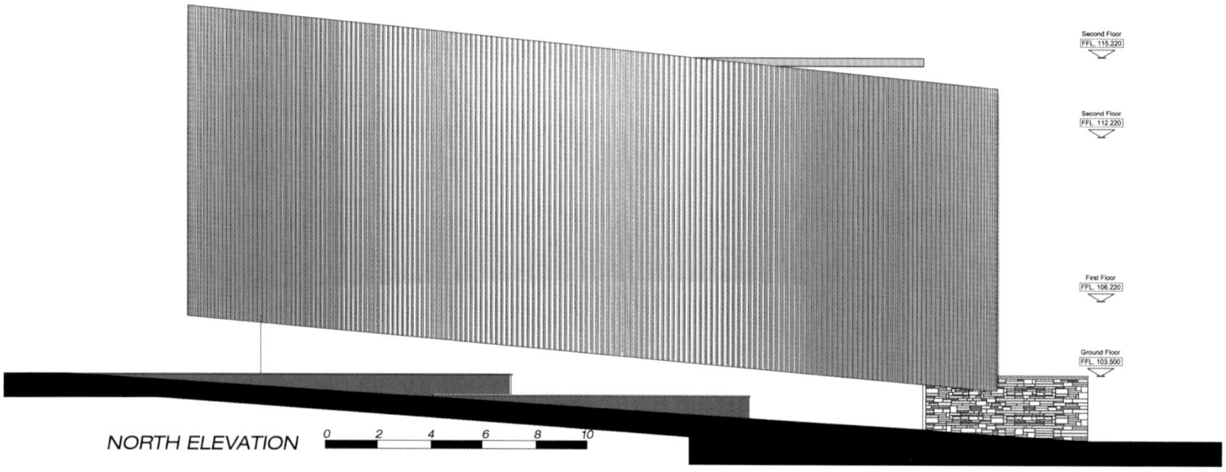

NORTH ELEVATION

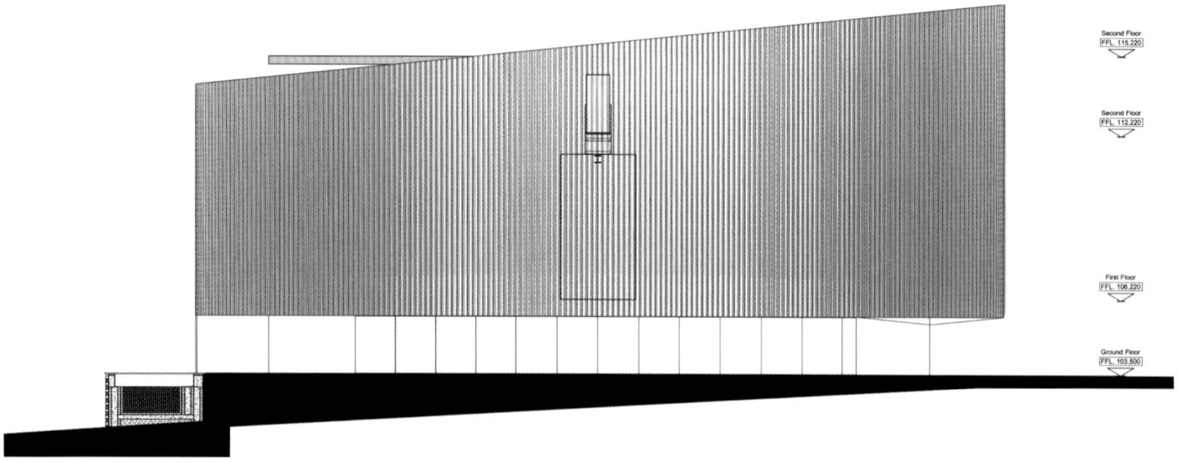

SOUTH ELEVATION

这个建筑的形式已经从一个狭窄场地的规限，变成了具有里程碑意义的雕塑。Fins和scrims创建成进出楼宇可视的联系，主画廊破茧而出。该设计是基于一个椭圆形的理念，并不代表一个令人惊喜的经典形象，而是有些东西有机地形成了。它反映了艺术的抽象性，引发人们的思考，从该建筑物的形式不难看出。正如其名称所示，它不是具体的定义，它是抽象的估量。

建筑的主要特点是铝片组合，分内部和外部空间，形成一个双层外立面，外面的第一层是网状分布的铝片。这些铝片沿着外墙重复布置创造了不朽的雕塑感建筑。作为元素，他们的力量在于光影沿着建筑的边缘漫射开来。随着时间和季节的推移不停变幻。阳光从北侧掠过，突出了空间感。幽默地表现诗情画意，颜色也变化无穷。此灵感来自于大自然，其隐含着秩序和混乱的矛盾。由于芦苇和草的运用在传统Zulu Kraals中很普遍，其线条优雅，这些元素通过外立面展现使之得以保持这种理念。

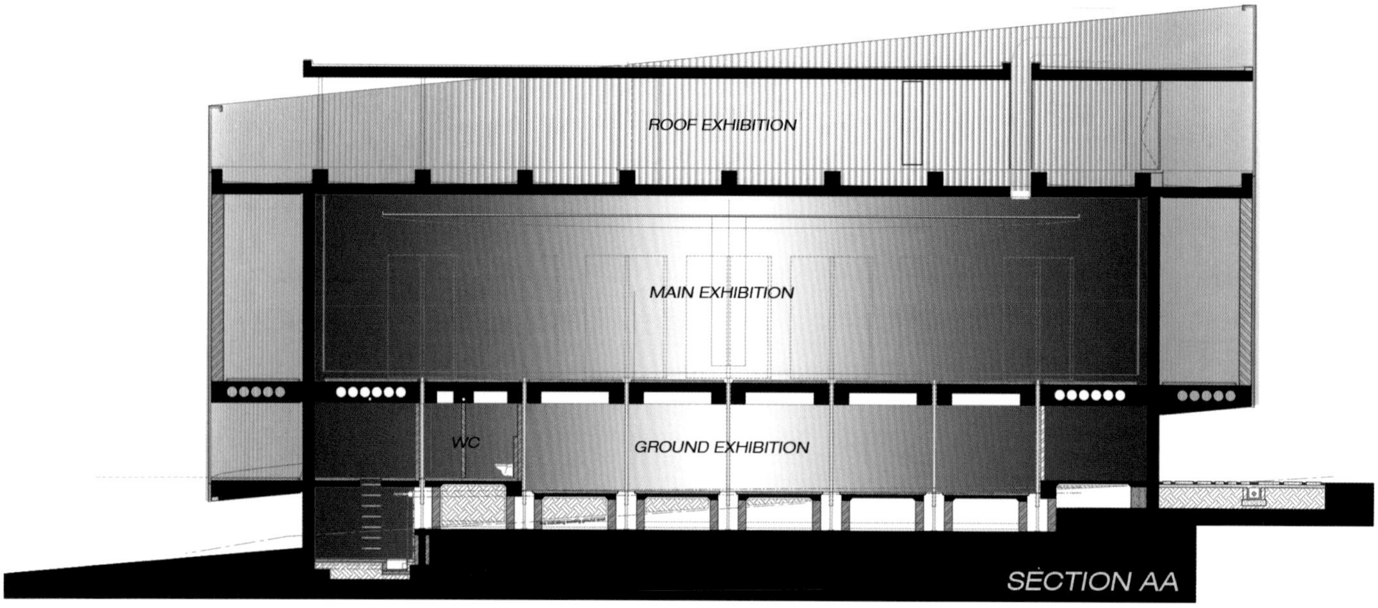

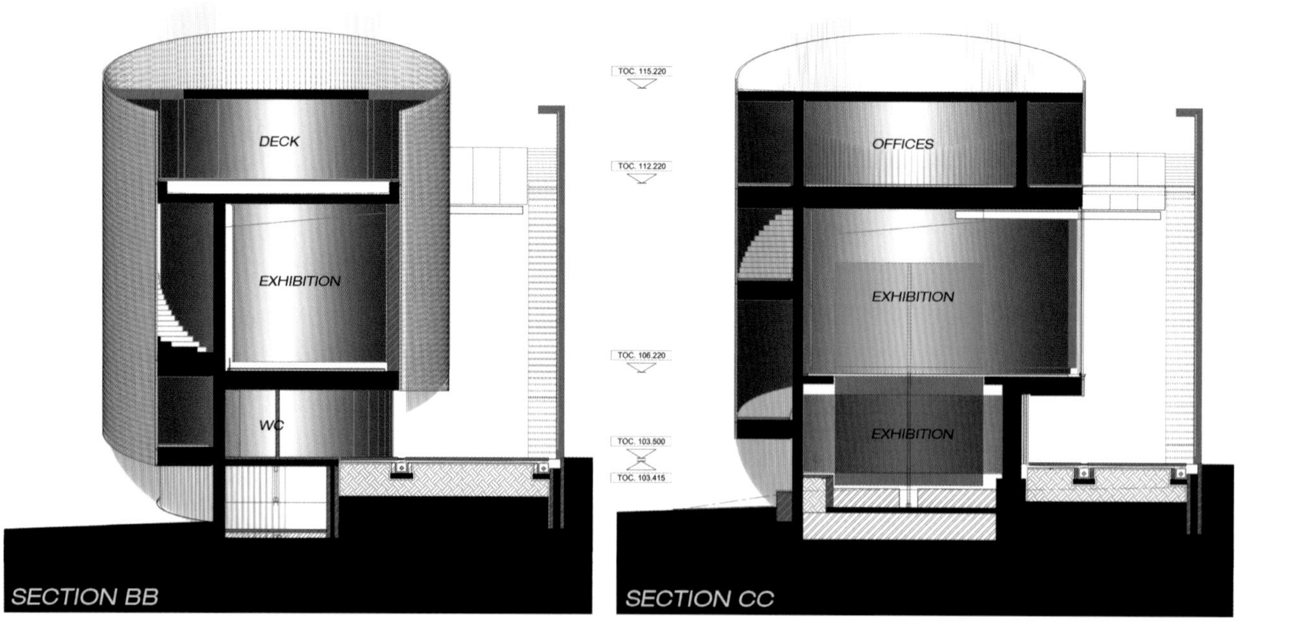

楼梯通道位于框架结构内，独立于主体建筑。通过线网的分布，使攀缘植物能沿着它生长，并覆盖住它。这是一段形式感极强，如同鳍状立起的结构。这件雕琢的产品（外观）和自然生长（绿色网格），交相辉映的目的是为了唤起在我们的城市，赞赏诗意的美好记忆。

项目为三个楼层。底层是为斯皮克和工艺品展览会提供的有关空间，面积约106m^2。登上公共区域的门槛，将用户吸引到旁边的建筑里，通过坡道连接地下及一楼展览空间，来促使人们经过排列整齐的外立面内侧，进入中心以后看到双卷的第一层，有177.76m^{2}多功能展览空间，有7个移动的显示器，可以通过地板看到地面以下的景象，实现了多功能楼全貌屏幕展示。顶层包括休息室和一个面积约85m^{2}的私人空间，功能上还能留出一个20m^{2}的露台，俯瞰约翰内斯堡。V

CIRCA
Art does not reproduce the visible; rather it makes it visible. - Paul Klee
Opened by Valli Moosa 3 November 2009

特别鸣谢：

中川国际
ZHONGCHUAN LNTERNATIONAL

中川国际建设（北京）有限公司建筑装饰设计研究院

公司创建于2004年，是一家综合性的建筑装饰装修企业，已通过ISO9001国际质量体系认证、ISO14001环境管理体系认证、GB/T28001职业健康安全管理体系认证，是中国建筑装饰协会会员单位。

经典案例
中国人民银行研究生部招待所及学生公寓改造工程
泛海国际居住区5#-8#地块精装修工程
泛海国际居住区5#-8#地会所精装修工程
国际港
光彩国际高级公寓精装修工程
兴隆家园及渝舍配套
观湖国际2号写字楼装饰工程
金融世家精装修工程
鄂尔多斯市邦城房地产开发有限责任公司的样板间和售楼处工程
中海广场香港能源办公楼精装修工程... ...

地址：北京亦庄经济技术开发区荣昌东街甲5号隆盛大厦B座5层
邮编：100176
电话：86-10-67806960/6965
传真：86-10-67806648
邮箱：Bjjsmhzs@163.com
网址Http://www.bjjsmhzs.com

Add: 5th Floor, Tower B, Longsheng Building, No.5 Rongchang East Street, Yi Zhuang Economic & Deve-lopment District Beijing, China Zipcode:100176
(Tel):86-10-67806960/6965
Fax:86-10-67806648
Email:Bjjsmhzs@163.com